ASSESSMENT OF MEDICINAL PLANTS FOR HUMAN HEALTH

Phytochemistry, Disease Management,
and Novel Applications

Innovations in Plant Science for Better Health: From Soil to Fork

ASSESSMENT OF MEDICINAL PLANTS FOR HUMAN HEALTH

Phytochemistry, Disease Management, and Novel Applications

Edited by
Megh R. Goyal
Durgesh Nandini Chauhan

APPLE
ACADEMIC
PRESS

Apple Academic Press Inc.
4164 Lakeshore Road
Burlington ON L7L 1A4
Canada

Apple Academic Press Inc.
1265 Goldenrod Circle NE
Palm Bay, Florida 32905
USA

First issued in paperback 2021

Library and Archives Canada Cataloguing in Publication

Title: Assessment of medicinal plants for human health : phytochemistry, disease management, and novel applications / edited by Megh R. Goyal, Durgesh Nandini Chauhan.

Names: Goyal, Megh Raj, editor. | Chauhan, Durgesh Nandini, editor.

Series: Innovations in plant science for better health.

Description: Series statement: Innovations in plant science for better health : from soil to fork | Includes bibliographical references and index.

Identifiers: Canadiana (print) 20200256491 | Canadiana (ebook) 20200256718 | ISBN 9781771888578 (hardcover) | ISBN 9780429328541 (ebook)

Subjects: Materia medica, Vegetable. | LCSH: Medicinal plants. | LCSH: Phytochemicals.

Classification: LCC RS164 .A87 2021 | DDC 615.3/21—dc23

Library of Congress Cataloging-in-Publication Data

Names: Goyal, Megh Raj, editor. | Chauhan, Durgesh Nandini, editor.

Title: Assessment of medicinal plants for human health : phytochemistry, disease management, and novel applications / edited by Megh R. Goyal, Durgesh Nandini Chauhan.

Other titles: Innovations in plant science for better health.

Description: Palm Bay, Florida, USA : Apple Academic Press, 2021. | Series: Innovations in plant science for better health: from soil to fork | Includes bibliographical references and index. | Summary: "This volume looks at the importance of medicinal plants and their potential benefits for human health, providing insight with scientific evidence on the use of functional foods in the treatment and management of certain diseases. Divided into four sections, the volume covers the assessment and identification of medicinal plants, the role of medicinal plants in disease management, the ethnobotany and phytochemistry of medicinal plants, and novel applications of plants. Chapters discuss the variety of bioactive compounds (also referred to as phytochemicals) in the leaves, stems, flowers, and fruits of certain plants that can help to promote human health. It outlines the bioactive molecules that can be isolated from medicinal plants, the available sources, the biochemistry and structural composition of the plants, and their potential biological activities. It goes on to look at bioactive compounds in relation to their potential pharmacological applications in human health, particularly disease prevention and management. Assessment of Medicinal Plants for Human Health: Phytochemistry, Disease Management, and Novel Applications sheds light on the potential of certain plants and will be of value to faculty and advanced-level students of natural products, food science, pharmacognosy, pharmacology, and biochemistry. It will also be of interest to researchers in the area of drug discovery and development. Key features: Looks at the preservation of indigenous knowledge on functional foods and medicine Discusses how to identify novel plants and their bioactive compounds for the treatment of different diseases Describes the mechanisms of action of the plants Encourages the development of plant-based drugs Fosters the integration of herbal medicine and modern medicine"-- Provided by publisher.

Identifiers: LCCN 2020023864 (print) | LCCN 2020023865 (ebook) | ISBN 9781771888578 (hardcover) | ISBN 9780429328541 (ebook)

Subjects: LCSH: Materia medica, Vegetable. | Medicinal plants. | Phytochemicals.

Classification: LCC RS164 .A85 2021 (print) | LCC RS164 (ebook) | DDC 615.3/21--dc23

LC record available at https://lccn.loc.gov/2020023864

LC ebook record available at https://lccn.loc.gov/2020023865

Apple Academic Press also publishes its books in a variety of electronic formats. Some content that appears in print may not be available in electronic format. For information about Apple Academic Press products, visit our website at **www.appleacademicpress.com** and the CRC Press website at **www.crcpress.com**

OTHER BOOKS ON PLANT SCIENCE FOR BETTER HEALTH BY APPLE ACADEMIC PRESS, INC.

Book Series: *Innovations in Plant Science for Better Health: From Soil to Fork*
Editor-in-Chief: Hafiz Ansar Rasul Suleria, PhD

Assessment of Medicinal Plants for Human Health: Phytochemistry, Disease Management, and Novel Applications
Editors: Megh R. Goyal, PhD, and Durgesh Nandini Chauhan, MPharm

Bioactive Compounds of Medicinal Plants: Properties and Potential for Human Health
Editors: Megh R. Goyal, PhD, and Ademola O. Ayeleso

Bioactive Compounds from Plant Origin: Extraction, Applications, and Potential Health Claims
Editors: Hafiz Ansar Rasul Suleria, PhD, and Colin Barrow, PhD

Cereals and Cereal-Based Foods: Functional Benefits and Technological Advances for Nutrition and Healthcare
Editors: Megh Goyal, PhD, Kamaljit Kaur, PhD, and Jaspreet Kaur, PhD

Health Benefits of Secondary Phytocompounds from Plant and Marine Sources
Editors: Hafiz Ansar Rasul Suleria, PhD, and Megh Goyal, PhD

Human Health Benefits of Plant Bioactive Compounds: Potentials and Prospects
Editors: Megh R. Goyal, PhD, and Hafiz Ansar Rasul Suleria, PhD

Plant- and Marine-Based Phytochemicals for Human Health: Attributes, Potential, and Use
Editors: Megh R. Goyal, PhD, and Durgesh Nandini Chauhan, MPharm

Plant-Based Functional Foods and Phytochemicals: From Traditional Knowledge to Present Innovation
Editors: Megh R. Goyal, PhD, Arijit Nath, PhD, and Hafiz Ansar Rasul Suleria, PhD

Plant Secondary Metabolites for Human Health: Extraction of Bioactive Compounds
Editors: Megh R. Goyal, PhD, P. P. Joy, PhD, and Hafiz Ansar Rasul Suleria, PhD

Phytochemicals from Medicinal Plants: Scope, Applications, and Potential Health Claims
Editors: Hafiz Ansar Rasul Suleria, PhD, Megh R. Goyal, PhD, and
Masood Sadiq Butt, PhD

The Role of Phytoconstitutents in Health Care: Biocompounds in Medicinal Plants
Editors: Megh R. Goyal, PhD, Hafiz Ansar Rasul Suleria, PhD,
and Ramasamy Harikrishnan, PhD

The Therapeutic Properties of Medicinal Plants: Health-Rejuvenating Bioactive Compounds of Native Flora
Editors: Megh R. Goyal, PhD, PE, Hafiz Ansar Rasul Suleria, PhD, Ademola Olabode
Ayeleso, PhD, T. Jesse Joel, and Sujogya Kumar Panda

ABOUT THE SENIOR EDITOR-IN-CHIEF

Megh R. Goyal, PhD

Retired Professor in Agricultural and Biomedical Engineering, University of Puerto Rico, Mayaguez Campus; Senior Acquisitions Editor, Biomedical Engineering and Agricultural Science, Apple Academic Press, Inc.

Megh R. Goyal, PhD, PE, is a Retired Professor in Agricultural and Biomedical Engineering from the General Engineering Department in the College of Engineering at the University of Puerto Rico–Mayaguez Campus; and Senior Acquisitions Editor and Senior Technical Editor-in-Chief in Agriculture and Biomedical Engineering for Apple Academic Press, Inc. He has worked as a Soil Conservation Inspector and as a Research Assistant at Haryana Agricultural University and Ohio State University.

During his professional career of 49 years, Dr. Goyal has received many prestigious awards and honors. He was the first agricultural engineer to receive the professional license in Agricultural Engineering in 1986 from the College of Engineers and Surveyors of Puerto Rico. In 2005, he was proclaimed as "Father of Irrigation Engineering in Puerto Rico for the Twentieth Century" by the American Society of Agricultural and Biological Engineers (ASABE), Puerto Rico Section, for his pioneering work on micro irrigation, evapotranspiration, agroclimatology, and soil and water engineering. The Water Technology Centre of Tamil Nadu Agricultural University in Coimbatore, India, recognized Dr. Goyal as one of the experts "who rendered meritorious service for the development of micro irrigation sector in India" by bestowing the Award of Outstanding Contribution in Micro Irrigation. This award was presented to Dr. Goyal during the inaugural session of the National Congress on "New Challenges and Advances in Sustainable Micro Irrigation" held at Tamil Nadu Agricultural University. Dr. Goyal received the Netafim Award for Advancements in Microirrigation: 2018 from the American Society of Agricultural Engineers at the ASABE International Meeting in August 2018.

A prolific author and editor, he has written more than 200 journal articles and textbooks and has edited over 80 books. He is the editor of three book series published by Apple Academic Press: Innovations in Agricultural & Biological Engineering, Innovations and Challenges in Micro Irrigation, and Research Advances in Sustainable Micro Irrigation. He is also instrumental in the development of the new book series Innovations in Plant Science for Better Health: From Soil to Fork.

Dr. Goyal received his BSc degree in engineering from Punjab Agricultural University, Ludhiana, India; his MSc and PhD degrees from Ohio State University, Columbus; and his Master of Divinity degree from Puerto Rico Evangelical Seminary, Hato Rey, Puerto Rico, USA.

ABOUT THE EDITOR

 Durgesh Nandini Chauhan, MPharm, has several years of academic (teaching) experience at institutes in India in pharmaceutical sciences. She taught subjects such as pharmaceutics, pharmacognosy, traditional concepts of medicinal plants, drug delivery, phytochemistry, cosmetic technology, pharmaceutical engineering, pharmaceutical packaging, quality assurance, dosage form designing, and anatomy and physiology. Presently she is working at the Ishita Research Organization, Raipur, India as a freelance writer and is guiding pharmacy, Ayurvedic and science students in their research projects.

She is a member of the Association of Pharmaceutical Teachers of India, SILAE: Società Italo-Latinoamericana di Etnomedicina (The Scientific Network on Ethnomedicine, Italy), and others. Her previous research work included "Penetration Enhancement Studies on Organogel of Oxytetracycline HCL." She has attended several workshops, conferences, and symposiums, including the AICTE-Sponsored Staff Development Program on "Effects of Teaching and Learning Skills in Pharmacy: Tool for Improvement of Young Pharmacy Teachers." She has written more than ten articles published in national and international journals, 13 book chapters, and two books: *Optimization and Evaluation of an Organogel* and *Plant- and Marine- Based Phytochemicals for Human Health: Attributes, Potential, and Use* (Apple Academic Press). She is also active as a reviewer for several international scientific journals and an active participant at national and international conferences, including Bhartiya Vigyan Sammelan and the International Convention of the Society of Pharmacognosy.

Mrs. Chauhan earned her BPharm degree in pharmacy from the Rajiv Gandhi Proudyogiki Vishwavidyalaya, Bhopal, India, and her MPharm (Pharmaceutical Sciences) in pharmaceutics from Uttar Pradesh Technical University (currently Dr. A.P.J. Abdul Kalam Technical University), Lucknow, India.

CONTENTS

CONTRIBUTORS

Ashwani Kumar Bhardwaj, PhD
Defence Institute of High Altitude Research (DIHAR), Ministry of Defence, Govt. of India,
c/o 56 APO, Leh-Ladakh 901205, Jammu and Kashmir, India
Plant Tissue Culture Laboratory, Neva Plantation Pvt. Ltd., Village Gopalpur 176059,
Dist. Kangra, HP, India. E-mail: ashwanibhardwaj67@gmail.com

Pooja Bohra, PhD
Division of Horticulture and Forestry, ICAR-Central Island Agricultural Research Institute,
Port Blair- 744105, Andaman and Nicobar Islands, India. E-mail: pooja.bohra24@gmail.com

Om Prakash Chaurasia, PhD
Defence Institute of High Altitude Research (DIHAR), Ministry of Defence, Govt. of India,
c/o 56 APO, Leh-Ladakh 901205, Jammu and Kashmir, India. E-mail: opchaurasia1998@gmail.com

Durgesh Nandini Chauhan
Assistant Professor, Columbia College of Pharmacy, Raipur, Chhattisgarh, India.
E-mail: pharmanandini@gmail.com

C. Donmez, PhD
Gazi University, Faculty of Pharmacy, Department of Pharmacognosy, Ankara, Turkey.
E-mail: ceylanaka@gazi.edu.tr

G. D. Durbilmez, PhD
Gazi University, Faculty of Pharmacy, Department of Pharmacognosy, Ankara, Turkey.
E-mail: goksendilsatdurbilmez@gmail.com

H. EL-Seedi, PhD
Uppsala University, Division of Pharmacognosy, Department of Medicinal Chemistry, Box 574,
SE-75 123, Uppsala, Sweden. E-mail: Hesham.ElSeedi@fkog.uu.se

Armando Enrique González-Stuart, PhD
School of Pharmacy, University of Texas at El Paso (UTEP),1101 N. Campbell St., Suite 110,
El Paso 79902, Texas, USA. E-mail: asgonzalez1@utep.edu

Megh R. Goyal, PhD
Retired Faculty in Agricultural and Biomedical Engineering from College of Engineering at University
of Puerto Rico–Mayaguez Campus; and Senior Technical Editor-in-Chief in Agricultural and
Biomedical Engineering for Apple Academic Press Inc.; PO Box 86, Rincon–PR–006770086, USA;
E-mail: goyalmegh@gmail.com

Yogini S. Jaiswal, PhD
Center for Excellence in Post-Harvest Technologies, The North Carolina Research Campus,
Kannapolis 28081, NC, USA. E-mail:yoginijaiswal@gmail.com

Chinthamani Jayavel, MSc
Alagappa University,Karaikudi 630003, Tamil Nadu, India. E-mail: jchinthu333@gmail.com

Saravanan Kandasamy, PhD
Division of Fisheries Science, ICAR-Central Island Agricultural Research Institute, Port Blair 744105,
Andaman and Nicobar Islands, India. E-mail: sarocife@gmail.com

Sahil Kapoor, MSc, PhD
Defence Institute of High Altitude Research (DIHAR), Ministry of Defence, Govt. of India,
c/o 56 APO, Leh-Ladakh 901205, Jammu and Kashmir, India. E-mail: skpuindia@gmail.com

T. R. Prashith-Kekuda, PhD
Department of Microbiology, S.R.N.M.N College of Applied Sciences, N.E.S Campus,
Balraj Urs Road, Shivamogga 577201, Karnataka, India. E-mail: p.kekuda@gmail.com

Navneet Kishore, PhD
Lab No A017, Department of Chemistry, University of Delhi, North Campus, Delhi 110007, India.
E-mail: kishore.navneet6@gmail.com

U. Koca-Caliskan, PhD
Gazi University, Faculty of Pharmacy, Department of Pharmacognosy, Ankara, Turkey.
E-mail: ukoca@gazi.edu.tr

Dhiraj Kumar, PhD
School of Studies in Zoology, Jiwaji University, Gwalior 475005, M.P., India.
E-mail: drkumarindia@yahoo.com

Kushal Kumar, PhD
Department of Pharmacology, ISF College of Pharmacy, Moga 142001, Punjab, India.
E-mail: kushal1kumar@gmail.com

Rosa Lelyana, PhD
Diponegoro University, Tampomas Dalam St VI No. 9, Semarang, Central Java, Indonesia.
E-mail: rl3lyana@gmail.com

Harishkumar Madhyastha, PhD
Department of Applied Physiology, Faculty of Medicine, University of Miyazaki, Kihara 5200,
Kiyotakecho, Miyazaki 889-1692, Japan. E-mail: hkumar@med.miyazaki-u.ac.jp

Radha Madhyastha, PhD
Department of Applied Physiology, Faculty of Medicine, University of Miyazaki, Kihara 5200,
Kiyotakecho, Miyazaki, 889-1692, Japan. E-mail: radharao@med.miyazaki-u.ac.jp

Vidyasagar Gunagambhere Manikrao, PhD
Medicinal Plants and Microbiology Research Laboratory, Department of Post Graduate Studies and
Research in Botany, Gulbarga University, Gulbarga 585106, Karnataka, India.
E-mail: gm.vidyasagar@rediffmail.com

Masugi Maruyama, PhD
Department of Applied Physiology, Faculty of Medicine, University of Miyazaki, Kihara 5200,
Kiyotakecho, Miyazaki, 889-1692 Japan. E-mail: masugi@med.miyazaki-u.ac.jp

Yuichi Nakajima, PhD
Department of Applied Physiology, Faculty of Medicine, University of Miyazaki, Kihara 5200,
Kiyotakecho, Miyazaki 889-1692, Japan. E-mail: yunakaji@med.miyazaki-u.ac.jp

Avilekh Naryal, PhD
Defence Institute of High Altitude Research (DIHAR), Ministry of Defence, Govt. of India,
c/o 56 APO, Leh-Ladakh 901205, Jammu and Kashmir, India. E-mail: aavilekh@gmail.com

Kanti Bhooshan Pandey, DPhil
CSIR-Central Salt & Marine Chemicals Research Institute, G. B. Marg, Bhavnagar 364002,
Gujarat, India. E-mail: kantibiochem@gmail.com

Alka Pawar, PhD
Dr. B. R. Ambedkar Center for Biomedical Research, University of Delhi, Delhi 110007, India.
E-mail: pawaralka18@gmail.com

Yutthana Pengjam, PhD
Faculty of Medical Technology, Prince of Songkla University, Hat Yai, Songkhla, Thailand.
E-mail: yutthana.p@psu.ac.th

Pavankumar Pindi, PhD
Department of Microbiology, Palamuru University, Mahabubnagar 509001, Telangana State, India.
E-mail: pavankumarpindi@gmail.com

Shivakumar Singh Policepatel, PhD
Department of Botany, Palamuru University, Mahabubnagar 509001, Telangana State, India.
E-mail: shivakumarsinghp@gmail.com

José O. Rivera, Pharm.D.
School of Pharmacy, University of Texas (UTEP), 1101 N. Campbell, Suite 110, El Paso 79902,
Texas, USA. E-mail: jrivera@utep.edu

Yatendra Kumar Satija, PhD
Dr. B. R. Ambedkar Center for Biomedical Research, University of Delhi, Delhi 110007, India.
E-mail: yksatija@gmail.com

Neetu Sharma, PhD
Krishi Vigyan Kendra (KVK) Kangra, CSKHPKV, Palampur 176001, District Kangra, H.P., India.
E-mail: neepradu@gmail.com

Brahm Kumar Tiwari, PhD
I.T.S Institute of Health & Allied Sciences, Murad Nagar, Ghaziabad 201206, U.P., India.
E-mail: brahmbiochem@gmail.com

K. S. Vinayaka, PhD
Department of Botany, Kumadvathi First Grade College, Shikaripura-577427, Shivamogga,
Karnataka, India. E-mail: ks.vinayaka@gmail.com

Ajit Arun Waman, PhD
Division of Horticulture and Forestry, ICAR-Central Island Agricultural Research Institute,
Port Blair- 744105, Andaman and Nicobar Islands, India. E-mail: ajit.hort595@gmail.com

Ashish Rambhau Warghat, PhD
CSIR-Institute of Himalayan Bioresource Technology, Palampur, Himachal Pradesh.
E-mail: aryanwarghat2810@gmail.com

Leonard L. Williams, PhD
Center for Excellence in Post-Harvest Technologies, The North Carolina Research Campus,
Kannapolis 28081, NC, USA. E-mail: llw@ncat.edu

ABBREVIATIONS

ABTS	2,2'-azino-bis(3-ethylbenzothiazoline-6-sulphonic acid)
AD	Alzheimer's disease
AGEs	advanced glycation end products
AMPK	adenosine monophosphate-activated protein kinase
ANI	Andaman and Nicobar Islands of India
AP-1	activator protein-1
AR	androgen receptor
ATCC	American type culture collection
ATP	adenosine triphosphate
Aβ	amyloid beta polypeptide
BCG	Bacille-Calmette Guerin
Bcl2	B-cell lymphoma 2
Bcl-xL	B-cell lymphoma-extra large
CAM	complementary or alternative medicine
CDK1	cyclin dependent kinase 1
CDKs	cyclin-dependent kinases
CFDA	China Food and Drug Administration
CFU	colony fluent unit
Chl	chlorophyll
CIARI	Central Island Agricultural Research Institute
CLRD	chronic lower respiratory diseases
COPD	chronic obstructive pulmonary disease
COX2	cyclooxygenase2
CREB	cAMP response element-binding protein
CRL	cerulenin
DCS	D-cycloserine
DDAs	diester diterpene alkaloids
Ddn	deazaflavin-dependent nitroreductase
DEDA	de-esterified diterpenoid alkaloid
DM	diabetes mellitus
DMF	dimethyl formamide
DPPH	2,2-diphenyl-1-picrylhydrazyl
DSHEA	Dietary Supplement and Health Education Act
ECM	extracellular matrix components

EGCG	epigallocatechin gallate
EGCG	epigallocatechin gallate
EMB	ethambutol
eNOS	endothelial nitric oxide synthase
EPTB	extra-pulmonary TB
ER	estrogen receptor
ERK	extracellular signal-regulated kinase
FDA	Food and Drug Administration
FOXO3a	forkhead box O3a
GADD	growth arrest and DNA damage inducible protein
GPx	glutathione peroxidase
GSK-3β	glycogen synthase kinase-3 beta
GST1	glutathione S-transferase 1
GULT	glucose transporter
HA	hyaluronic acid
HDL-c	high-density lipoprotein cholesterol
HeLa cells	human cell line by Henrietta Lacks
Her2/ Neu	Human epidermal growth factor receptor 2
HIF-1	hypoxia inducible factor-1
HIV	human immunodeficiency virus
HPLC	high performance liquid chromatography
HPU	Herbarium of Palamuru University
HSV	herpes simplex virus
HSV-1	herpes simplex virus type 1
IC	inhibitory concentration
ICAM	intercellular adhesion molecule
ICAR	Indian Council of Agricultural Research
ICL	isocitratelyase
IL	interleukin
INH	isoniazid
INH	isonicotinylhydrazide
iNOS	inducible nitric oxide synthase
ITK	indigenous technical knowledge
IUCN	International Union for Conservation of Nature
LDL-c	lipoprotein cholesterol
LOX-1	lipooxygenase-1
LTB	latent TB
M. acuminate	*Musa acuminate*
M. paradisiaca	*Musa paradisiaca*

MAPK	mitogen-activated protein kinase
MBC	minimum bactericidal concentration
MDA's	monoester diterpene alkaloids
MFC	minimum fungicidal concentration
mg/ml	milligram per milliliter
MIC	minimum inhibitory concentration
MJP	mulberry juice purification
MMP	mulberry marc purification
MMP	matrix metalloproteinase
MMP-2	matrix metalloproteinase-2
MMP-9	matrix metalloproteinase-9
MnSOD	manganese superoxidedismutase
MRSA	methicillin resistant *Staphylococcus aureus*
MS	mass spectrometry
MTB	mycobacterium tuberculosis
MTM	mexican traditional medicine
mTOR	mammalian target of rapamycin
MTT	3-(4, 5-dimethylthiazolyl-2)-2,5-diphenyltetrazolium bromide
NA	nutrient agar
NADH	nicotinamide adenine dinucleotide
NFκB	nuclear factor- kappa B
NFT	neurofibrillar tangles
NF-κB	nuclear factor kappa B
NMR	nuclear magnetic resonance
nNOS	neuronal nitric oxide synthase
NOS	nitric oxide synthase
NRF2	nuclear factor erythroid 2-related factor
PAHO	Pan American Health Organization
PAS	para-amino salicylate
PDA	potato dextrose agar
PDE	phosphodiesterase
PI3K	phosphoinositide 3-kinase
PKC	protein kinase C
PTB	pulmonary TB
PZA	pyrazinamide
RIF	rifampicin
RNA	ribonucleic acid
ROS	reactive oxygen species
RSV	respiratory syncytial virus

SGLT	sodium-glucose transport proteins
T1D	type 1 diabetes
T2D	type 2 diabetes
TB	tuberculosis
TCM	traditional Chinese medicine
TCN	trans-cinnamic acid
TDR	totally drug-resistant
TH	traditional healers
TK	traditional knowledge
TLC	thin layer chromatography
TNF	tumor necrosis factor
TSM	traditional systems of medicine
v/v	volume by volume
VCAM	vascular cell adhesion molecule
VCAP	vascular cell adhesion protein
WHO	World Health Organization
XDR	extensively drug-resistant

PREFACE 1

Prologue

Benefits of Amla (*Phyllanthusemblica*): *Slows down ageing.* *Cures heart/jaundice diseases;* *and sore throat.* *Increases diuretic and metabolic activities;* *and hair growth.* *Prevents ulcers, constipation, graying of hair,* *formation of GB stones.* *Makes skin glow.* —**Jashandeep Kaur** 11th grade student [*Souce*: *Roshni* (light). In: *Official Annual Book*; Dhuri (Punjab): Government Girls Senior Secondary School; Volume 8; 2019; page 64]	*To be healthy is our moral responsibility,* *towards Almighty God, ourselves,* *our family, and our society; and no life without nature.* *Eating fruits and vegetables makes us healthy,* *believe and have a faith.* *Reduction of food waste can reduce the world hunger and can make our planet eco-friendly.* —**Megh R. Goyal** Lover of Mother Nature Senior Editor-in-Chief

Medicinal plants contain certain chemicals in their organs such as leaves, stem, root and fruits that can provide therapeutic benefits against different kinds of diseases. These chemicals are often referred to as "phytochemicals." The word "phyto" is a Greek word that means "plant." Phytochemicals are natural nonessential bioactive compounds found in plants/plant foods. Thousands of phytochemicals have already been identified, and more are still being discovered every year. Plants that are used for medicinal purposes are often considered to be less toxic and induce fewer side effects than synthetic medicine. In our world today, many commercially available drugs have plant-based origins, with more than 30% of modern medicines directly or indirectly derived from medicinal plants. Indeed, plants can be a major source of pharmaceutical agents in the treatment of many life-threatening diseases.

The use of medicinal plants has largely increased because they are locally accessible, economical, and as well vital in promoting health. However, scientific data and information regarding the safety and efficacy of these medicinal plants are inadequate. There is an awareness to use plant-based products as is evident from the short health tips from *Phyllanthusemblica* at the beginning of this preface.

We introduce this book volume under book series *Innovations in Plant Science for Better Health: From Soil to Fork.* This book mainly covers the current scenario of the research and case studies and contains scientific evidence on the health benefits that can be derived from medicinal plants and how their efficacies can be improved. The findings reported in this book can be useful in health policy decisions and will also motivate the development of health care products from plants for health benefits. The book will further encourage the preservation of traditional medical knowledge of medicinal plants. Therefore, these plant products are drawing attention of researchers and policymakers because of their demonstrated beneficial effects against diseases with high global burdens such as diabetes, hypertension, cancer, and neurodegenerative diseases.

This book volume is a treasure house of information and an excellent reference for researchers, scientists, students, growers, traders, processors, industries, dieticians, medical practitioners, users, and others. We hope that this compendium will be useful for students and researchers as well as those working in the food, nutraceutical, and herbal industries.

The contributions by the cooperating authors to this book volume have been most valuable in the compilation. Their names are mentioned in each chapter and in the list of contributors. We appreciate you all for having patience with our editorial skills. This book would not have been written without the valuable cooperation of these investigators, many of whom are renowned scientists who have worked in the field of plant science and food science throughout their professional career.

The goal of this book volume is to guide the world science community on how plant-based secondary metabolites can alleviate us from various conditions and diseases.

We will like to thank Apple Academic Press, for making every effort to publish this book when all are concerned with health issues. Special thanks are due to the AAP production staff for typesetting the entire manuscript and for the quality production of this book.

We request the reader to offer your constructive suggestions that may help to improve the next edition.

We express our admiration to our families and colleagues for understanding and collaboration during the preparation of this book volume. As an educator, there is a piece of advice to one and all in the world:

"Permit that our almighty God, our Creator, provider of all and excellent Teacher, feed our life with Healthy Food Products and His Grace–; and Get married to your profession.—"

—Megh R. Goyal, PhD, PE
Senior Editor-in-Chief

PREFACE 2

The plan of writing a book entitled *Assessment of Medicinal Plants for Human Health: Phytochemistry, Disease Management, and Novel Applications* is to draw attention to the advancement of medicinal plants in phytochemistry and their therapeutic applications. The book is divided into four parts: Part I—Assessment and identification of medicinal plants; Part II—Ethnobotany and phytochemistry of medicinal plants; Part III—Role of medicinal plants in disease management; and Part IV—Novel applications of plants.

First two chapters are dedicated to the ethnomedicinal use of plants followed by chapters on mulberry, rhodiola, piper and aconite to spotlight the pharmacological importance.

The chapters on the role of phytochemicals in combating diabetes, hemorrhoids, and tuberculosis are also included. Bioactive benefits of banana, coffee, and lichen are also covered. The review chapters focus on plants used as hyaluronidase and gelatinase inhibitors and importance of the nutraceuticals in translational medicine.

This book bridges an important gap and would be useful for academicians, experts in ethanomedicine, scientists, and students perusing education in phytomedicine, health and natural product, as well as to herbal industries.

We appreciate the sincere efforts by Dr. Kamal Shah, Associate Professor, GLA University Mathura, India, for his expertise in preparing this book.

Finally, I would like to thank family members for allowing me to take time away from their precious company. To my colleague Megh R. Goyal, my husband Dr. Nagendra. S. Chauhan (Drugs Testing Laboratory, Government of India, Raipur), and our daughters Harshita and Ishita, for their inspiration, motivation and guidance.

—**Durgesh Nandini Chauhan, MPharm**
Editor

PART I

Assessment and Identification
of Medicinal Plants

CHAPTER 1

ASSESSMENT OF ETHNOMEDICINAL PLANTS: SHIWALIK HILLS, HIMACHAL PRADESH, INDIA

NEETU SHARMA

ABSTRACT

From time immemorial, medicinal plants have played crucial role in the primary health care system for the natives of the study area and still today they are practicing herbal remedies. Due to the rough topography, lack of western pharmaceuticals and inappropriate health services, the local people have strong cultural belief regarding the benefits of folk medicines. Medicinal plant species are increasingly under threat due to destructive harvesting especially from wild and this may lead to extinction of species in future. Together with increasing population, increased demands of the pharmaceutical market for a continuous and uniform supply of crude medicines and lack of a basic knowledge base, it is not possible to assure the continued availability of plant material from the wild. Thus, there is a need to generate awareness among local communities toward the sustainable utilization and conservation. The best conservation planning strategy can only be accomplished by coordinated involvement of traditional healers (TH), government, and all the organizations. TH have sufficient knowledge of sustainable harvesting, which should be practiced and employed in the conservation practices, while the government agencies should manage and control different stakeholders dealing with conservation processes.

1.1 INTRODUCTION

India, a land of enormous biotic wealth, has more than 7000 plant species, which have been reported to be used for herbal medicines and most of them are being exploited for the extraction of drugs. During the recent years, the indigenous uses of medicinal plants have acquired significance for discovering newer and more effective plant-based drugs. Plants still remain the basis for development of modern drugs.[2] Medicinal plants are considered as rich natural resource, which can be utilized in drug development. The biochemical components present in the plants are considered to have compatibility with the human body.[12] Joy et. al.[11] reported about 120 therapeutic agents of known structure from different parts of about 90 species. Podophyllotoxin, gitoxigenin, pilocarpine, vinblastine, curcumin, vincristine, aspirin, and ephedrine are widely known examples of such useful plant-based drugs. In Asia, India is one of the prime countries, where vast wealth of traditional knowledge (TK) system prevails till today.

Himalayas is a rich depository of medicinal plant wealth and traditional medicinal knowledge. Due to its unique climate and topographic diversity, it is endowed with variety of habitats, which can improve the possibility for the growth of potential medicinal plant species. In India, Himachal Pradesh (HP) state is popularly known for use of the plant wealth in indigenous knowledge system and is a veritable paradise for diversity of plant species.[17] HP with wider deviations in the altitude ranging from the plain areas (such as Una, Hamirpur) to higher mountains (Kinnaur and Lahaul Spiti districts). Because of erratic altitudes, there is significant variation in temperature and rainfall, cropping patterns, soil types, and flora and fauna of the state. However, being a hilly state, the region is less accessible yet has sustained good wealth of TK on preparation of herbal drugs, which has been sporadically recorded in the past and there is still ample knowledge that needs to be documented. In every community of the state, the information on the traditional use of medicinal plants with health benefits is passed on orally from generation to generation. They rely on medicinal plants because these are cost effective, safe, and with least side effects. Local poor communities of the state are using herbal medicines to fight against incurable diseases.

Inaccessibility to modern health care, the rising costs of modern medicines (mostly imported) coupled with serious side effects of certain drugs have re-emerged the need of traditional plant-based medicines.[5] Interest in herbal medicine is being fueled by wider acceptability and the documentation of rich TK about medicinal plant species is gaining urgent priority to revive local health traditions and technologies[1,4]. Traditionally, local communities in the world possess unique knowledge on the local plant resources on which they still rely because traditional medicines are efficacious and are so intimately and immediately dependent.

Shiwalik region of HP is no exception as it has also been very rich in such traditional heritage. Unfortunately, much of the accumulated extraordinarily abundant indigenous knowledge of medicinal plants—which have been acquired due to their long-term practices and handed down usually orally, from one generation to another—is dwindling because of the loss of the traditional culture and changes in sustainence economy.[7,8] Hence, proper documentation of these traditional practices is the need of the hour and the present study is an attempt to document important ethnomedicinal formulations practiced for treating various ailments by inhabitants of Hamirpur and Kangra regions in HP. Thus, to preserve the traditional system of herbal medicine and the process of discovering many new drugs, it is highly important to document TK on ethnomedicinal use of such medicinal plants.[13] Number of studies have well acknowledged[6,14] that the use of herbal plants to treat various diseases leads to further development of many life saving modern drugs.

This chapter focuses on the assessment of ethnomedicinal use of medicinal plants in Shiwalik hills of HP in India.

1.2   MATERIALS AND METHODS

HP is situated in the North West Himalayas of Indian subcontinent and possesses unique heritage of ethnobotanical floral diversity. The HP state ranges from Shiwalik hills in the south to the greater Himalayas ranges in the north embracing an area of 55,538 km^2. The state lies between longitude 75°45'55" of Eastern side to 79°04'20" East and latitudes of 30°22'40" Northern to 33°12'40" North. Owing to hilly and rugged terrain in the entire state of HP, the altitude ranges from 350 m to 7000 m above sea. HP is categorized into three zones, such as:[9]

- **Outer Himalayas:** District Una, Hamirpur, Bilaspur, Kangra, and the lower parts of Mandi, Solan, and Sirmaur areas, where altitude varies from 350 m to 1500 m.
- In **the Inner Himalayas/mid-mountains,** the altitude ranges from 1500 m to 4500 m above mean sea level and comprises parts of upper regions of Renuka and Pachhadin of the Sirmaur district, tehsils of Mandi district Karsog and Chachiot Churah tehsil of Chamba district.
- The altitude of **Alpine zone or the greater Himalayas** is above 4500 m above mean sea level and includes area of Lahaul and Spiti district, areas of Kinnaur district and Pangi tehsil of Chamba district.

The study area in the district Hamirpur of HP has a latitude of 31°41'10.23" North and a longitude of 76°31'16.71" East. There are nearly 15,000 species of phanerogams or flowering plants in the district, but unfortunately only 17% have been recognized as possessing great potential of medicinal importance. Hamirpur district is characterized by wide phytodiversity of broad-leaved and deciduous plants and has been considered as absolute emporium of medicinal plants.[18]

The Kangra District is situated between the latitude of 32°5'59.29" North and a longitude of 76°16'8.76" East covers an area of 5739 km². It occupies about 30% of the geographical area under forests and the vegetation varying from alpine pastures at higher altitude to dry scrub forests at lower altitude to produce a rich botanical wealth and large number of diverse plants. These forests harbor wider plant genetic diversity that forms the skeleton system of the traditional health care system.

Resource persons such as traditional healers (TH), old experienced men and women, and many local informants with knowledge of medicinal plants and their efficacy were consulted to collect the data. The observations were based on local folklore and interview-based survey along with informal discussions and field visits. Interviews were conducted in the local dialect.

1.3   RESULTS AND DISCUSSION

Ethnobotanical explorations highlighted the medicinal use of 48 plant species, which belong to 30 families in the Kangra District. Table 1.1 presents species name, family name, local name, purpose of using a particular plant, parts of plant used and detailed information pertaining to mode of herb administration. Highest representative of species under study belonged to family *Liliaceae* followed by *Amaranthaceae, Euphorbiaceae, Fabaceae, Lamiaceae, Rutaceae, Solanaceae followed by Leguminosae, Asclepiadaceae, Moracea, Combretaceae*; and besides these *Solanaceae* and *Poaceae, Liliaceae, Araceae, Euphorbiaceae, Polygonaceae, Ranunculaceae, Scrophularaceae and Valerianaceae; and Verbenaceae, Apocynaceae, Meliaceae, Moraceae, Pinaceae, Plantaginaceae, Rutaceae, Apiaceae, Saxiferagaceae, and Zingiberaceae* and other 20 families were used by local natives for medicinal use.

Qualitative analysis pertaining to parts of plants used, methods of preparation, and administrative routes and applications revealed that the TH in the study area are utilizing different parts of medicinal plants to prepare various herbal drugs for curing various human ailments/diseases (Table 1.1). Amongst plant parts used, most commonly used part was leaf

TABLE 1.1 Medicinal Plants Used for Curing Various Diseases/Ailments.

Plant name (*Family)	Local name	Purpose of use	Part used	Mode of administration
Acacia catechu Willd. *Mimosaceae	Khair	Wounds	Resin	Aqueous extract is applied externally till healing
		Cough	Root	Decoction of dried root is prescribed
		Postpartum care	Whole plant	Decoction prepared by boiling katha along with Elletaria Cardamomum L. is served to woman after 2–3 days of delivery to provide strength to the body and also to help in secretion of milk. To reduce body pain after delivery, the water boiled with heartwood chips of Acacia catechu Willd is used to take bath by woman.
		Mouth sores	Resinz	Resin is directly applied on the sores.
		Rheumatism	Root	Paste of fresh root is put on the joints once a day for a week.
		Diarrhea	Bark	Decoction of 15 g of bark in 1 L of water is served orally at regular intervals.
Acacia nilotica (L.) Willd.ExDelile *Leguminosae	Kikar	Prevention of uterus prolapse	Bark	Sittings are done in water boiled with bark (200 g each) of kikar and Ficusreligiosa L.
Acyranthesaspera L. *Amaranthaceae	Puth-kanda	Dry cough	Seed	Powder obtained from roasted seeds is mixed with honey and served with luke warm water before going to bed.
		Blisters of mouth	Leaf	Leaf juice provides relief.
		Ease in delivery	Root	Root paste is put evenly over the navel and pubic region.
		Stomach ache	Root	Decoction of roots is advised.
Acoruscalamus L. *Acoraceae	Barein	Chest congestion; cough	Rhizome	Rhizome paste is applied on chest to cure chest congestion and cough.
		Fever	Rhizome	Rhizome paste along with warm mustard oil is helpful.

TABLE 1.1 *(Continued)*

Plant name (*Family)	Local name	Purpose of use	Part used	Mode of administration
		Arthritis	Rhizome	Rhizomepasteis applied on the joint with pain.
		Wound	Leaf	External application of leaf paste is suggested.
		Stomach trouble	Root	Extract or decoction of root in small dosage can relieve stomach trouble.
		Toothache	Root	Chewing the root alleviates toothache.
		Abortion	Root	An infusion of the root is served.
Aeglemarmelos L. Corr. *Bael* Rutaceae*		Diarrheaand dysentery	Fruit	One fourth portion of an unripe bael is consumed orally.
		Toothache	Fruit	Filtered mixture of unripe fruit boiled in water is used for gargling.
		Intestinal disorder	Fruit	Squash prepared from fruit pulp added to boiled waster is served.
		Constipation	Fruit	Half portion of a ripe fruit is taken two times a day up to 3–4 days.
Ajugabracteosa Wall. exBenth. *Lamiaceae	*Neel kanthi*	Mouth ulcer and throat infection	Leaf	Leavesof *Ajugabracteosa* along with leaves of *Centellaasiatica* are chewed.
Allium cepa L. *Liliaceae	*Piaz*	Boils	Bulb	Paste of bulb is applied externally on the affected portion.
		Postdelivery pain	Bulb	Bulbs boiled with mustard oil are messaged on the affected area.
		Morning sickness	Bulb	Mixture of crushed *Alliumcepa* L. and few leaves of *Menthaspicata* L. with a pinch of salt is taken orally.
Allium sativum L. *Liliaceae	*Lahsun*	Postdelivery pain	Bulb	Bulbs boiled with mustard oil are messaged on the affected area.
		Blemishes on face	Bulb	Paste prepared from ground *Allium sativum* L. and unripe *Carica papaya* L. is applied on face.
Aloe barbadense Mill. *Liliaceae	*Kavarein*	Burns	Leaf	Extracted gel or pulp of leaves is applied on skin.

TABLE 1.1 *(Continued)*

Plant name (*Family)	Local name	Purpose of use	Part used	Mode of administration
		Pain during menstruation	Leaf	About 15 mL extract of leaf is taken once for 2–3 days.
		Sun stroke	Leaf	Pulp extracted from the leaves is served orally.
		Eye infection	Leaf	After peeling the skin of leaf, the pulpy matter is bandaged on eyes for an overnight.
		Constipation	Leaf	Leaves are ground and paste is prepared by adding salt and turmeric to it. Half teaspoon of this paste is served.
Amaranthus Caudatus L. *Amarnthaceae	*Chulai*	Intestinal inflammation	Leaf	Fresh leaf juice along with curd is consumed thrice a day.
Asparagus racemosus L. *Liliaceae	*Satawari*	Urinary trouble	Root	About 4–5 teaspoonful of dried root powder is given thrice a day for a week
		Postdelivery weakness	Root	Root decoction is served thrice a day up to fornight.
		Excessive bleeding after delivery	Root	Fresh tuberous root juice (15 mL) is served after meal for 2 days.
		Fertility	Root	Equal quantities of roots *of Asparagus* and *Bambaxceiba* L. Along with seeds of *Piper nigrum* L. are ground and paste is prepared. Two spoons of paste is mixed in goat milk; and is administered daily before breakfast from third day of menstruation upto 5 days.
		Liver disorder; enhance lactation	Root	Dried root powder is served.
		Improve stamina	Root	Dried root powder mixed with boiled water and cow's milk is served to the mother.
Azadirachtaindica A.Juss. *Meliaceae	*Neem*	Irritation in eyes	Leaf	Leaves are boiled in water. After cooling, it is used to wash the eyes.
		Acne/pimples	Leaf	Ground paste of leaves of neem is applied on acne, till it is dried out.

TABLE 1.1   (Continued)

Plant name (*Family)	Local name	Purpose of use	Part used	Mode of administration
		Wounds	Leaf	Leaf paste is dabbed on wounds a few times till healing.
		Dandruff	Leaf	Leaves are boiled in water and are allowed to cool. After washing the hair with shampoo, cleanse them with leaf boiled water.
		Skin problems	Leaf	Turmeric added to paste of neem leaves can be used for eczema and ring worms.
		Fever	Leaf	Leaf Juice is administered.
		Dental care	Twigs	Twigs are used to clean teeth.
Bauhinia variegate L. *Fabaceae	*Kachnar/ Karalen*	Snake bite	Root	Root juice is served.
		Skin disease	Bark	Decoction of bark is prescribed.
		Prolonged menstruation	Flower bud	Decoction of flower buds is given.
		Sore throat	Bark	Gargling with decoction prepared from bark is prescribed.
		Promote milk secretion	Flower	Flower decoction isserved to woman as a galactogogue.
Butea monosperma (Lamk.) Taub. *Fabaceae	*Dhak/Palash*	Dysentery	Resin	Resin is taken orally.
		Pimples	Bark	Bark is applied as poultice.
		Expulsion of germs	Seed	Seed powder is given.
		Wounds	Bark	External application of bark paste on wounds is suggested.
Calotropisprocera (Ait). R. Br. * Asclepiadaceae	*Aak*	Abortion	Leaf	During initial stages of pregnancy, leaf is slightly saute in castor oil and is placed on vagina for whole night.

TABLE 1.1 *(Continued)*

Plant name (*Family)	Local name	Purpose of use	Part used	Mode of administration
		Backache during pregnancy	Leaf	Warm yellow leaves smeared with mustard oil are placed on the back.
		Sprain	Leaf	Turmeric and onion combined with crushed leaves are bandaged on the affected part.
		Boils	Leaf	Slightly heated leaves smeared with seasame oil are placed on boils.
Cannabis sativa L. *Cannabaceae	Bhang	Insect bite	Leaf	Crushed leaves are rubbed to cure insect bite.
		Cough and cold	Seed	Roasted seed powder is served with boiled water.
		Wounds and bone fracture	Leaf	Leaf extract is applied on wounds and fractured bones.
Calotropis gigantia (L.) R.Br. *Asclepiadaceae	Aak	Cough	Root	Burnt root is made into fine powder and one teaspoon of powder is taken in the morning and evening with lukewarm water.
		Scabies	Latex	Fresh latex is topically applied on the affected area two times a day for week.
		Arthritis	Leaf	Leaf juice mixed with sesame oil is massaged on affected area.
Carica papaya L. *Caricaceae	Papita	Abortion	Fruit and latex	Tender fruit is consumed two times a day for 5 days.
		Indigestion	Fruit	Pulp of ripe papaya fruit aids in digestion.
		Skin disease	Latex	The latex is applied topically to cure ringworm and itching of skin.
		Pimples	Fruit	Paste from pulp of raw papaya is applied as face mask.
Cassia angustifolia Vahl. *Leguminosae	Sanai	Constipation during pregnancy	Leaf	Leaves are chewed.
		Constipation	Leaf	About 2 g of dried leaf powder with lukewarm water is administered.

TABLE 1.1 *(Continued)*

Plant name (*Family)	Local name	Purpose of use	Part used	Mode of administration
Cassia fistula L. *Caesalpiniaceae	Gurlakdi	Skin disease	Leaf	Leaf paste in combination with vinegar is applied to the affected part.
		Constipation during pregnancy	Pod (pulp)	Boiled and strained mixture of one teaspoon of ground *Trachyspermumammi* (L.) Sprague, 2 cm piece of *Cassia fistula* L. and 1 teaspoon of ground *Foeniculum vulgare* Mill. is mixed in a glass of water; and served twice a day.
		Fever	Root	About 40–50 g of root are crushed in one cup of water and administered two times a day till recovery.
		Whooping cough	Fruit	Ash of burnt fruit is given.
		Cough	Fruit	Skin of fruit is chewed in the morning.
Chenopodium alba L. *Amaranthaceae	Bathu	Constipation	Leaf	Strained mixture of leaves boiled in water is combined with 1 teaspoon of lemon juice. One glass is consumed to cure constipation.
		Jaundice	Leaf	Chenopodium alba leaves cooked with equal quantity of spinach and any other leafy vegetable is taken.
		Diarrhea	Leaf	Soak dried leaves in water overnight. Boil it next morning. Mix 1 tsp of sugar and strain the liquid. Drink 1 tsp.
		Rheumatism	Leaf	About 15 g of leaf juice is taken empty stomach daily for 2–3 months. After drinking the juice, nothing should be consumed for about 2–3 hours.
Cuscutareflexa Roxb. *Convolvulaceae	Akashbel	Abortion	Stem	During early stages of pregnancy, paste (1 teaspoonful) of stem and lime is administered once in the morning for 5 days.
		Skin infection	Whole plant	Bath of water boiled with whole plant is prescribed.
		Nasal bleeding	Vine	Extract obtained after crushing the vine is poured into the nostrils.
		Constipation, flatulence	Vine	About 40–50 mL of vine decoction is served once a day.

TABLE 1.1 (Continued)

Plant name (*Family)	Local name	Purpose of use	Part used	Mode of administration
		Diarrhea	Vine	Decoction of vine is recommended.
		Scabies	Vine	Fresh paste of vine is applied topically over the affected area.
Cynodondactylon (L.) Pers. *Poaceae	Doob	Lice problem	Stem	Stem juice is used to kill lice.
		Nose bleeding	Leaf	Leaf juice is dropped into the nostrils.
		Cuts and wounds	Grass	Paste prepared from grass and garlic is bandaged on the affected part.
		Headache	Grass	Tea prepared from few grass leaves relieves headache.
CyperusrotundusL. *Cyperaceae	Naga-rmoth	Fever	Tuber	About 1 teaspoon of powdered tuber is given twice a day.
		Promote milk flow	Tuber	Fresh tuber paste is put on breasts of lactating mother.
		Conjunctivitis	Root	Root juice is dropped into eyes.
Foeniculum vulgare Mill. *Apiaceaea	Meethi-saunf	Morning sickness	Seed	One teaspoonful roasted and ground mixture of Foeniculum vulgare (50 g), mishri, and Trachyspermum ammi (10 g) is served with lukewarm water daily.
Ficus benghalensis L. *Moracea	Barh	Foot cracks	Latex	White latex is applied twice a day till healing.
		Diarrhea	Bark	Concoction prepared by boiling bark in water is served two times day for 1 week.
		Excessive bleeding	Leaf	One tablespoon of dried leaf powder helps in controlling excessive bleeding.
		Boils	Leaf	Slightly heated leaf smeared with deshi ghee is bandaged over the affected part.
		Cracked heels	Latex	Milky juice is applied on cracked heels.
Ficus religiosa L.*Moraceae	Pipal	Scrofula	Root	Paste prepared by crushing the root is applied topically.
		Gum problem	Roots	Roots are chewed.

TABLE 1.1 (Continued)

Plant name (*Family)	Local name	Purpose of use	Part used	Mode of administration
Jatropha curcus L. *Euphorbiaceae	*Ratanjot*	Abortion	Root	Root is introduced into vagina and is kept overnight for 2–3 days to induce abortion.
Lawsoniainermis L. *Lythraceae	*Mehndi*	Leukemia	Latex	Latex mixed with water is taken orally.
		Abortion	Root	Root decoction used in abortion.
		Sore throat	Leaf	Gargling with leaf decoction is prescribed.
		Burning sensation in the feet	Leaf	Leaf paste is applied on the soles of the feet.
		Dysentery	Bark	About 30–50 g of bark decoction is administered.
		Headache	Leaf	Paste prepared from dried leaves and water is applied on the forehead.
		Boils	Leaf	Fresh leaf paste is applied on the affected area.
Murrayakoenigii L. Spreng. *Rutaceae	*Gandala*	Swelling of feet and legs	Fresh leaf	Luke warm decoction is applied topically on swollen feet and legs.
		Dental problem	Stem	Stem is used as tooth brush.
		Dysentry	Leaf	Green leaves are chewed.
		Burns	Leaf	Paste fresh leaves is applied on burns.
		Vomiting	Leaf	The infusion of dried leaves is recommended to stop vomiting.
		Stomach pain	Leaf	Paste of few leaves mixed with a cup of buttermilk is taken in the morning on empty stomach.
Musa paradisiacal L. *Musaceae	*Kela*	Antifertility	Root	Juice of root is used orally to stop conception.
Ocimumgratissimum L *Lamiaceae.*	*Bhabri*	Abdominal pain during pregnancy	Leaf	Decoction of 3 g *Ocimumgratissimum* L., few leaves of *Mentha spicata* L., and 2 teaspoons of *Trachyspermum ammi* (L.) Sprague is consumed.

TABLE 1.1 (Continued)

Plant name (*Family)	Local name	Purpose of use	Part used	Mode of administration
Rhododendron arboretum Smith *Ericaceae	Barah	Diarrhea	Seed	Seeds are taken with milk.
		Diarrhea	Flower	Flower squash is given to cure diarrhea.
Ricinus communis L. *Euphorbiaceae	Aerand	Headache	Leaf	Leaf paste is applied on forehead.
		Nasal bleeding	Flower	Sniffed to stop nasal bleeding.
		Wounds	Leaf	Leaf poultice is applied on wounds.
		Joint pains	Fresh leaf	Luke warm oil is smeared on leaves and is applied topically for joint pains.
Sesamumorientale L. *Pedaliaceae	Til	Postdelivery pains	Seed	Seed oil is messaged for postdelivery pain.
		Amenorrhea	Seed	Seed powder is taken with milk.
Sida Cordifolia L. *Malvaceae	Daridein	Postdelivery pains	Seed	Seed powder roasted with butter is used to cure postdelivery pain and to improve vigor.
		Bleeding piles	Root	Root infusion (2 teaspoons) is served for 3 days.
Solanum nigrum *Solanaceae	Makoya	Painful joints	Leaf; stem	Slightly heated juice of leaves and stems are applied on painful joints once a day.
		Cough (children)	Fruit	Ripe fruit is served thrice a day for a week.
		Headache	Fruit	Juice of fresh fruit is applied over forehead.
		Conjunctivitis	Leaf	Leaf extract is kept over eyes 2–3 times a day.
Solanum surattense Burm. F. *Solanaceae	Kankari	Cough	Fruit, leaf	Decoction of leaves and dried fruit (10 g) is consumed twice for 4–5 days.
		Ear pain	Root	Few drops of root extract is poured into the ear.
		Tonsillitis	Whole plant	Plant extract is used as gargle.
Taraxacum Officinale F. H. Wigg. aggr. *Asteraceae	Dudli	Constipation during pregnancy	Leaf	Vegetable prepared from Taraxacum officinale and Spinacia oleracea L. leaves is given.

TABLE 1.1 (Continued)

Plant name (*Family)	Local name	Purpose of use	Part used	Mode of administration
Terminalia Bellirica (Gaertn.) Roxb. *Combretaceae	*Bahera*	Cough	Fruit	About 1 g of fruit powder is given with honey till recovery.
Terminalia chebula Retz. *Combretaceae	*Harad*	Indigestion	Fruit	The powder of dried fruits is advised.
		Constipation during pregnancy	Fruit	Decoction of one fruit each of *Terminalia chebula* Retz and *Terminalia bellirica* Roxb. and two fruits of *Emblica officinalis* Geartn. is administered once a day till problem persists.
		Constipation	Fruit	Fruit powder is consumed to cure constipation.
		Stomatitis.	Bark	Bark powder along with water is used for gargling.
Tinospora cordifolia (Willd.) *Menispermaceae	*Galoein*	Ease in delivery	Stem	Stem juice is given at the time of delivery.
		Increase in lactation	Stem	Decoction of stem (2") is prepared and is given half in morning and half in evening for 7–14 days.
		Fever	Stem	Stem decoction (1 teaspoon) is prescribed three times a day.
		Cough and cold	Stem	Mixture prepared by mixing stem powder with honey is given
Trigonellafoenum-graecum L. *Fabaceae	*Methi*	Backache during pregnancy	Seed	*Trigonellafoenum graecum* L. seeds (2 teaspoons) boiled in a glass of milk are given at bed time.
		Diabetes	Seed	15–20 g of seeds are taken orally for a fortnight.
Viola serpens Wall. *Violacae	*Bana-fshah*	Cold and cough during pregnancy	Vine	Two teaspoons decoction of *Viola serpens* Wall, ground *Elletaria Cardamomum* L., and *Glycyrrhiza glabra* L. is given to pregnant woman.
		Cough	Flower	Flower paste is consumed.
		Constipation	Root	Crushed root is given.

TABLE 1.1 (Continued)

Plant name (*Family)	Local name	Purpose of use	Part used	Mode of administration
Vitex Nigundo L. *Verbenaceae	*Bana*	Postpartum care	Leaf, seed	Leaves and seeds boiled in water are used for bathing new born babies and mothers.
		Joint pain	Root	Root decoction is applied on painful joints.
		Ear pain	Leaf	Mustard oil heated with leaves of *vitex negundo* is used as ear drop.
Withania Somnifera L. *Solanaceae	*Aas-gandh*	Increase lactation	Seed	Ground mixture of *Withania somnifera* (20 g), *Asparagus racemosus* (5 g), and *Glycyrrhiza globra* (1″) is given to lactating mother along with milk in morning and evening for 15 days.
		General weakness	Root	Root powder (10 g) along with milk is taken twice a day.
		Wounds	Root	Root paste is applied over the affected area.
Zanthoxylu malatum- Roxb. *Rutaceae	Tirmira	Throat irritation	Leaves and soft stem	Decoction used as gargle.
		Dental pain	Twigs	Twigs are used as tooth brush to avoid dental problem.
		Mouth ulcer	Leaf	Paste of leaves applied to cure mouth ulcer.
Zingiber Officinale Rosc *Zingiberaceae	*Adrak*	Backache during pregnancy	Rhizome	Four teaspoons of mixture of roasted and crushed dried ginger along with ghee and sugar is given daily. Ginger juice mixed with honey is served once a day upto 1 week.

(32%), followed by root (16.6%), fruits (9.5%), seeds (7.1%), stem (5.9%), bark (5.3%), latex (3.6%). Rhizome, flower, bulb, and vines were least used (3% each) while whole plant and resin were rarely used (1.8%). A meager percentage (1.2%) of grass, tuber, and twigs were used (Fig. 1.1).

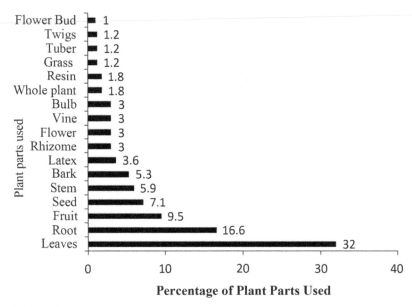

FIGURE 1.1   Statistical data on parts of plants for medicinal use.

Figure 1.1 showed that most preferred plants were chosen to prepare the herbal remedies because they contain more active phytoconstituents in comparison to flower bud, twigs, tuber, and grasses.[3] The present study revealed that leaf occupied the topmost rank in preparation of various herbal remedies that agrees with the results obtained from studies conducted earlier in HP,[15,19] where most of the medicinal plant species were harvested for utilization of their leaves and use of leaves may not cause much harm to the local plant diversity in the region in comparison to the plant species whose roots are utilized. However, on the contrary several other studies have shown that overharvesting of roots, stems, leaves, and flowers can negatively affect the survival of useful medicinal plant species and hence impacts sustainable exploitation of these medicinal herbal plants.[16]

As regards methods of preparation of herbal remedies, for preparing herbal drugs for different ailments, all parts of plant (such as leaf, seed, fruit, root, bark, etc.) have been used in different ways. The healers explained that selection of method for preparing a particular drug was based on some specific properties of the medicinal plants and the anticipated prepared drug. In this research, it was observed that among the various forms of preparations, paste, decoction, powder, and infusion prepared from different plant parts were the most commonly used.

Decoction is prepared by slow simmering of the less permeable and thicker part of the plant, for example, the roots, bark, fruit, and seed. Simmering is practiced for easy extraction of medicinal constituents. The plant material is chopped into smaller pieces and it is ascertained that the simmering pot is covered so as to conserve the volatile components of the decoction. The solids are separated from the liquid.

Infusion is a technique for extraction of biocompounds from plant parts dissolved in a solvent by allowing the material to remain suspended in the solvent. The solvent may be water, oil, or alcohol. In most of the infusions, the volatile components of the aerial parts, that is, roots, leaves, bark, flowers, fruits, etc, are extracted. For preparing a desired infusion, single part of the plant such as the root or a combination of other parts of the same plant, such as the leaves, bark, and the seed are required.

The method of using plants also varied according to the nature of the ailment. Decoction of plants was mostly prescribed for respiratory problems (cough, fever), digestive disorders (constipation, diarrhea, dysentery, stomachache), and maternal care (postdelivery weakness, milk secretion, abortion). For maternal health and care, similar results were observed in the prior surveys conducted on use of medicinal plants for maternity care in Hamirpur and Kangra districts of HP.[20] It was further revealed in this study that external application of the paste of juice/leaves of certain plants was suggested for treatment of various skin problems while ailments like boils, wounds, inflammation, etc. were treated by applying poultice of plant seeds/leaves/bulbs or by placing the leaves smeared with oil/deshi ghee. The dual treatment has been suggested for curing rheumatism: Leaf juice of *Chenopodiumalba* L. is administered internally while root paste of *Acacia catechu* Willd. is applied externally on the joints. The 31 remedies were based on pastes either to be applied externally (for joint pains, chest congestion, wounds, burns, skin diseases, blemishes on face, scabies, scrofula) or taken orally (constipation, stomachache, cough). For 24 ailments, decoctions were administered while powder from different plant parts was suggested for 18 ailments followed by infusion for 3 ailments. The other treatments included direct consumption (such as chewing) and external application etc. The plants for these ailments were used both as fresh or dried forms.

It was informed by the healers that these herbal drugs are being prepared either from combinations of numerous plants or by a single plant species. In total, 14 species—*Acacia nilotica, Ficusreligiosa, Allium sativum, Carica papaya, Terminalia chebula, Terminalia bellirica, Emblica officinalis, Cassia fistula, Ajuga bracteosa, Allium cepa, Ocimum gratissimum, Taraxacum officinale, Viola serpens, Withania somnifera*—were used in multiple plant-based remedies in combination with each other or with other plant species such as *Trachyspermum ammi, Foeniculum vulgare, Centella asiatica, Bambax ceiba, Piper nigrum, Mentha spicata, Spinacia oleracea, Elletaria cardamomum, Glycyrrhiza glabra,* and *Asparagus racemosus.*

The joint use of multiple medicinal plants may be due to synergetic or additive effects of two or more plants.[10] Traditional practitioners also believe that combination of medicinal plants for preparing a drug gives better results in comparison to using the single plant alone as the combination of multiple plants increases the quality and effectiveness of medicine. Inhabitants of Mustang in Nepal also recorded similar observations.[3]

Out of total 48 plant species, 32 plant species (i.e., *Acacia catechu, Acyranthes aspera, Acoruscalamus, Aegle marmelos, Allium cepa, Aloe barbadense, Asparagus Racemosus, Azadirachta indica, Bauhinia variegate, Butea monosperma, Calotropis procera, Cannabis sativa, Calotropis gigantia, Carica papaya, Cassia angustifolia, Cassia fistula, Chenopodium alba, Cuscuta reflexa, Cynodon dactylon, Cyperus rotundus, Ficus benghalensis, Lawsonia inermis, Murraya koenigii, Rhododendron arboretum, Solanum nigrum, Solanum surattense, Terminalia chebula, Tinospora cordifolia, Viola serpens, Vitex nigundo, Withania somnifera, Zanthoxylum alatum*) are used for treatment of more than two ailments. Thus, the importance of a particular plant increases with the increase in number of the diseases/ailments it can cure. It was found that *Acorus calamus, Azadirachta indica* and *Cuscuta reflexa* were used to treat maximum number of ailments, that is, seven each.

1.4 SUMMARY

Village elders have tremendous knowledge about medicinal plants and the need of hour is sustainable exploration, development, and use of their knowledge and plant wealth. Despite the tremendous influence and dependency on latest medication, rural population still rely on herbal medicines as they are readily available, safe, and cost effective. Due to advancements in modern technologies and gradual depletion of traditional culture, the indigenous technical knowledge (ITK) regarding therapies and medicinal plants of local natives have almost reached on the verge of extinction. Hence, with a view to preserve this cultural and traditional heritage of our local practitioners/ healers, the present study was conducted to gather and document the medicinal plants, which are traditionally being used for treating various ailments.

This study concluded that the rural people of the study area depend on crude preparations of nearly fifty plant species that are used singly or in combination to treat various ailments, that is, digestive problems, respiratory problems, skin problems, orthopediatric ailment, dental troubles, maternal health, etc. Almost all plant parts, that is, root, stem, bark, leaf, flower, tuber, seed, and fruit were used for making different preparations; and among all these leaf was most commonly used part followed by root, fruit, seeds, stem, bark, and latex. Tuber and twigs were rarely used. Most of the herbal medicines reported in this chapter were prepared in the form of paste, powder, decoction, infusion, etc. Infusion was principally employed on leaf/root extracts while decoction was used on fruit, leaf, roots, and bark extracts. It was note worthy to observe that some of the plants were observed to have the ability to cure a number of ailments whereas some of them were effective for a particular ailment. It is must to preserve our cultural heritage by ensuring adequate tools for sustainability and conservation of plants in botanical gardens.

KEYWORDS

- **ethnomedicine**
- **herbs**
- **Himachal Pradesh**
- **indigenous users**
- **traditional healers**

REFERENCES

1. Anim, A. K.; Laar, C.; Osei, J.; Odonkor, S.; Enti-Brown, S. Trace Metals Quality of Some Herbal Medicines Sold in Accra, Ghana. *Int. Acad. Ecol. Environ. Sci.* **2012**, *2* (2), 111–117.
2. Ates, D. A.; Erdogrul, O. T. Antimicrobial Activities of Various Medicinal and Commercial Plant Extracts. *Turkish J. Biol.* **2003**, *27*, 157–166.

3. Bhattarai,S.; Chaudhary, R. P.; Quave, C. L.; Taylor, R. S. The Use of Medicinal Plants in the Transhimalayan Arid Zone of Mustang District, Nepal. *J. Ethnobiol. Ethnomed.* **2010,** *6,* 14.

4. Budovsky, A.; Fraifeld, V. E. Medicinal Plants Growing in the Judea Region: Network Approach for Searching Potential Therapeutic Targets. *Netw. Biol.* **2012,** *2* (3), 84–94.

5. Chapman, K. R.; Chomchalow, N. Production of Medicinal Plants in Asia. In *Medicinal Plants Research in Asia, Volume 1: The Framework and Project Workplans*; Batugal, P. A., Kanniah, J., Lee, S. Y., Oliver, J. T. Eds.; International Plant Genetic Resources Institute—Regional Office for Asia, the Pacific and Oceania (IPGRI-APO): Serdang, Selangor DE, Malaysia, 2004; Vol. 1, pp 33–42.

6. Cox, P. A.; Ballick, M. The Ethnobotanical Approach to Drug Discovery. *Sci. Am.* **1994,** 270 (6), 82–87.

7. Gangwar, K. K.;D.; Gangwar, R. S. Ethnomedicinal Plant Diversity in Kumaun Himalaya of Uttarakhand, India. *Nat. Sci.* **2010,** *8* (5), 66–78.

8. Hamilton, A. The People and Plant Initiative: X–XI. In *Ethnobotany*; Martin, G. J. Ed.; Chapman and Hall: London. 1955; pp 3–7.

9. http://nidm.gov.in/PDF/DP/HIMACHAL.PDF (accessed January 31, 2019).

10. Igoli, J. O.; Ogaji, O. G.; Tor-Anyiin, T. A.; Igoli, N. P. Traditional Medicine Practice Amongst the Igede People of Nigeria, Part II. *Afr. J. Tradit. Complement. Altern. Med.* **2005,** *2* (2), 134–152.

11. Joy, P. P.; Thomas, J.; Mathew, S.; Skaria, B. P. Medicinal Plants. *Trop. Hortic.* **1998,** *2,* 449–632.

12. Kamboj, V. P. Herbal Medicine. *Curr. Sci.* **2000,** *78* (1), 35–39.

13. Khandel, A. K.; Ganguly, S.; Bajaj, A.; Khan. S. New Records, Ethno-Pharmacological Applications and Indigenous uses of *Gloriosasuperba* L. (Glory Lily) Practices by Tribes of Pachmarhi Biosphere Reserve, Madhya Pradesh, Central India. *Nat. Sci.* **2012,** *10* (5), 23–48.

14. Kirtikar, K. R.; Basu, B. D. Eds., *Indian Medicinal Plants with Illustrations*. 2nd ed.; Oriental Enterprises: Dehradun, India, 2001; p 839.

15. Kumar, S.; Kumar, P. Medicinal Plant Diversity in Tungal Valley of District Mandi, Himachal Pradesh (India). *Asian J. Adv. Basic Sci.* **2014,** *2* (3), 103–108.

16. Lulekal, E.; Kelbessa, E.; Bekele T.; Yineger, H. An Ethnobotanical Study of Medicinal Plants in Mana Angetic District, Southeastern Ethiopia. *J. Ethnobiol. Ethnomed.* **2008,** *4,* 10.

17. Sharma, A.; Gupta, A. K. Exploration of some Medicinal Plants used by Natives of District Hamirpur (H.P.). *Int. J. Scientific Res. Publ.* **2012,** *2* (7), 1–5.

18. Sharma, N.; Sharma, G. K. Exploration of some Ethnomedicinal Plants Used by Natives of Hamirpur District in Himachal Pradesh. *Asian Agri. Hist.* **2015,** *19 (2)*, 95–104.

19. Sharma, N. Application of Ethnomedicinal Plants in Traditional Healthcare Systems Prevalent in Shiwalik Hills, Himachal Pradesh, India. In *Recent Advances in Ethnobotany*; Kumar, S., Ed.; Deep Publications: New Delhi, 2015; pp 107–115.

20. Sharma, N. Ethnobotanical Survey of Medicinal Plants used for Maternity Care in Kangra and Hamirpur districts of Himachal Pradesh. In *Ethnobotanical Studies in India*; Kumar, S. Eds.; Deep Publications: New Delhi, 2014; pp 245–255.

CHAPTER 2

INDIGENOUS MEDICINAL PLANTS IN ACHAMPET FOREST: TELANGANA, INDIA

SHIVAKUMAR SINGH POLICEPATEL, PAVANKUMAR PINDI, and VIDYASAGAR GUNAGAMBHERE MANIKRAO

ABSTRACT

The wisdom of original 25 curative plants has been practiced to cure skin, jaundice, infertility, bone fracture, nephrology diseases by pastoral people of Telangana, India. The current targets on the importance of indigenous curative plants insight have been documented in this study. This information might contribute meticulously in recent drug manipulative or government policies to improve contemporary novel drug design systems in rural, folk-loric areas, and in the enrichment of advance formulas with reference to indigenous medicinal plants.

2.1 INTRODUCTION

The rural people still depend on indigenous plants for treatment of diverse diseases. These people have useful information on the biochemical properties of medicinal plants from their forefathers. They collect medicinal plants from nearby forests and use these plant materials as raw drugs. However, as a result of transformation and human's uncontrolled performance, lifestyle of these people have changed rapidly and ultimately resulting in loss of indigenous knowledge among rural folks. Therefore, efforts should be made to document various uses of these plants before some of these are eliminated from the natural habitats, or before these residents shift over to modern remedies. The indigenous people live near forests so that they possess huge wisdom on properties and use of medicinal plants.

Indigenous remedial medicine is a learning or evaluation of the conventional medicine practiced by ancestral groups, and particularly by aboriginal people. Indigenous medicine has been useful in the continued existence among humans and practically in all cultures throughout the world. The formulated indigenous remedy is a synonym for conventional medication. Indigenous medicine is cost effective, safe, and easily accessible.[5,6]

In this chapter, novel wisdom on indigenous medicinal plants have been documented and discussed to cure five major disorders related to skin, jaundice, infertility, bone fracture, and nephrology.

2.2 RESEARCH METHODOLOGY

Field trials and surveys were conducted in Achampet Forest of Nagarkurnool District of Telangana State, India. Each participation of innovative wisdom was documented yearly. The secret knowledge was collected based on discussions with the local practitioners and users (i.e., elders) from the selected areas.[1,3,4] The discussions were conducted several times with authenticated evidences that were collected and identified using local floras.

The reported documented area in Telangana state is located in South India, which is a region with great indigenous wisdom on medicinal plants to treat diverse diseases. The flora of Telangana covers the diversity of identifications. The present study focuses on the momentous aboriginal curative vegetations, which require to be recognized for various prospective uses.

2.3 RESULTS AND DISCUSSION

Indigenous remedial plants and their wisdom have been deep rooted in Indian aboriginal cultures. The wisdom of original curative plants is being still practiced. These statements are relevant to the ethnic acquaintance of indigenous medicinal plants that are exploited in this chapter to treat five major disorders by rural people of Telangana, India.

The 72 indigenous medicinal plants were identified to cure skin, jaundice, bone fracture, infertility, and kidney disorders by pastoral people of Achampet forest in Telangana state, India. The 28 species were documented on the rustic knowledge of unique plant-based natural remedies for skin diseases, compared to 21 species of normal therapeutic plants to treat jaundice. Total of nine species of indigenous medicinal plants belonging to eight families were documented to treat infertility. Additional 14 species of therapeutic plants belonging to 10 families were observed to treat kidney disorders. The current targets on the importance of indigenous curative plants were documented. The variety of wisdom velour on recent prescription manipulations with the novel contemporary novel improvement creates rural and ethnic locations,[2] with the enrichment of their advance prescription methods for indigenous curative therapeutic plants.

2.3.1 INDIGENOUS MEDICINAL PLANTS TO TREAT SKIN DISEASES

Common skin diseases (such as eczema, leukoderma, ringworm, scabies, and many erstwhile circumstances) can be treated with indigenous plant-based drugs. In India, there is a huge information base of indigenous plants to cure skin diseases.

The reachable information is rigorous on the rural wisdom of 28 species representing 19 families on original medicinal plants to cure skin diseases (Table 2.1). The details of individual species have been orderly determined. The present assortment of contemporary medicine devious in industrial policies to encroach novel medicine schemes in rustic origin vicinity includes enrichment of advance formulations with rural curative medicinal plants.

2.3.2 INDIGENOUS REMEDIAL PLANTS TO TREAT JAUNDICE

The 21 species concerned with 17 families have been documented as normal therapeutic plants to treat jaundice the rural area of Telangana, India (Table 2.2). This includes leading families Ceasalpiniaceae and Euphorbiaceae each with three species, followed by Fabaceae with two species and the remaining 14 families. In the survey, importance of indigenous medicinal plants in the therapeutic use of jaundice was detected. The combination of information might contribute meticulously in the improvement of contemporary medicine schemes, in pioneering prescription design systems in aboriginal locations.

TABLE 2.1 Indigenous Medicinal Plants for Treating Skin Diseases.

Botanical name	Family	Habitat	Local name	Part of plant used
Abutilon indicum	Malvaceae	Climber	Duvvena benda (Telugu)	Leaf
Acalypha indica	Euphorbiaceae	Herb	Maaredu (Telugu), Beel (Hindi)	Fruit
Achyranthes aspera	Amaranthaceae	Herb	Uttareni (Telugu), Aapang (Hindi)	Leaves
Adhatoda vasica	Acanthaceae	Shrub	Addasaramu (Telugu), Adoosa (Hindi)	Leaves
Annona squamosa	Annonaceae	Shrub	Seethaphalamu (Telugu), Seethaphal (Hindi)	Leaves
Argemone mexicana	Papavaraceae	Herb	Zeripothu alamu (Telugu)	Latex
Aristolochia bracteolata	Aristolochiaceae	Herb	Eeshhwari (Telugu)	Bulb
Azadirachta indica	Miliaceae	Tree	Veepa (Telugu), Neem (Hindi)	Leaves
Barleria prionitis	Acanthaceae	Tree	Velakkaya (Telugu), Kabeet (Hindi)	Fruit
Breynia vitisidaea	Euphorbiaceae	Climber	Madhu nashini (Telugu & Hindi)	Whole plant
Cassia fistula	Fabaceae	Herb	Gandham (Telugu), Chandan (Hindi)	Ripened leaves
Cassia accidentalis	Fabaceae	Herb	Yaknayk aaku (Telugu).	Leaves
Celosia argentea	Amaranthaceae	Shrub	Gumugu (Telugu), Kaale Jaamun (Hindi)	Leaves
Citrullus colocyanthis	Cucurbitaceae	Climber	Thippa teega (Telugu), Amrutha (Hindi)	Leaf
Clerodendrum inerme	Verbenaceae	Herb	Takkulapu chettu (Telugu), choti Aari (Hindi)	Flower
Daucus carota	Apiaceae	Herb	Karet (Telugu), Gajar (Hindi)	Leaves
Dillenia indica	Dilleniaceae	Tree	Kalinga (Telugu), Panchapaal (Hindi)	Flower
Diospyros montana	Ebenaceae	Tree	Kakaulmedu (Telugu), kaladhao (Hindi)	Leaves
Jatropha gossypifolia	Euphorbiaceae	Shrub	Adavi amudamu (Telugu), Arandi (Hindi)	Bark
Lepidogathis cristata	Acanthaceae	Shrub	Mullabanthi (Telugu), Bhukar zadi (Hindi)	Flower
Mucuna pruriens	Fabaceae	Tree	Durada Gondi (Telugu), Kooch (Hindi)	Leaves

TABLE 2.1 *(Continued)*

Botanical name	Family	Habitat	Local name	Part of plant used
Pongamia pinnata	Fabaceae	Tree	Kanugu (Telugu), Kaaranga (Hindi)	Seeds
Santalum album	Santalaceae	Tree	Chandanam (Telugu), Chandan (Hindi)	Leaves
Strycnos nuxvomica	Loganiaceae	Herb	Musti (Telugu), Khajra (Hindi)	Bark
Tamarindus indica	Fabaceae	Tree	Chintha (Telugu), Emli (Hindi)	Leaves
Thevetia nerrifolia	Apocynacae	Shrub	Ganneru (Telugu & Hindi)	Fruit
Wrightia tinctoria	Apocynaceae	Shrub	Ankudu (Telugu), Kapar (Hindi)	Leaves

TABLE 2.2 Indigenous Medicinal Plants for Treating Jaundice.

Botanical name	Family	Local name	Habitat	Part of plant used
Abrus precatorius.	Fabaceae	Guruginja (Telugu), Gunchi (Hindi)	Climber	Leaves
Acacia arabica	Mimosaceae	Thumma (Telugu), Babul (Hindi)	Tree	Fresh bark peel
Aloe barbadensis MILL.	Liliaceae	Kalabanda (Telugu), Ghikanvar (Hindi)	Herb	Arial part
Argemone mexicana L.	Papaveraceae	Zeeripothu allamu (Telugu), Bharbandh (Hindi)	Shrub	Seeds
Aristolochia bracteolata LAM.	Aristolochiaceae	Gaadede (Telugu), Ausala (Hindi)	Herb	Bulb
Balanites aegyptiaca (L.) Delile	Balanitaceae	Gaara (Telugu), Baam (Hindi)	Tree	Fruit
Caesalpinia bonduc (L.) Roxb.	Ceasalpiniaceae	Gajja kaya (Telugu), Gajaga (Hindi)	Shrub	Leaves
Caesalpinia sappan L.	Ceasalpiniaceae	Baakanu chekka (Telugu), Baakam (Hindi)	Shrub	Bark
Calotropis gigantea (L.) R.BR.	Asclepiadaceae	Zilledu (Telugu), Aakan (Hindi)	Herb	Roots
Cassia fistula L.	Ceasalpiniaceae	Argyadamu (Telugu), Aalis (Hindi)	Herb	Leaves
Holarrhena antidy-senterica (ROTH.) A.DC.	Apocynaceae	Paalakodisa (Telugu), Dhuudi haat (Hindi)	Tree	Leaves
Lawsonia inermis L.	Lythraceae	Mydaaku (Telugu), Mahandi (Hindi)	Shrub	Leaves
Leucas aspera (WILLD.) SPRENG.	Lamiaceae	Thummi (Telugu), Chota halkusa (Hindi)	Shrub	Leaves
Oroxylum indicum (L.)VENT.	Bignoniaceous	Aaku maanu (Telugu), Arlu (Hindi)	Shrub	Seeds
Phyllanthus fraternus Webster	Euphorbiaceae	Neela usiri (Telugu), Hazaramani (Hindi)	Herb	Leaves
Ricinus communis L.	Euphorbiaceae	Aaku maanu (Telugu), Arlu (Hindi)	Shrub	Leaves
Solanum nigrum L.	Solanaceae	Kashabusha (Telugu), Chirpoti (Hindi)	Herb	Leaves
Tephrosia purpurea PERS.	Fabaceae	Vempalle (Telugu), Sarapunkha (Hindi)	Herb	Leaves
Terminalia chebula RETZ.	Comrataceae	Karkkaya (Telugu), Balhar (Hindi)	Tree	Fruit
Tinospora cordifolia (WILLD.)	Minispermaceae	Thippa theega (Telugu), Adharvela (Hindi)	Climber	Leaves
Vernonia cinerea LESS.	Asteraceae	Garita kammi (Telugu), Sahadevi (Hindi)	Herb	Seeds

2.3.3   INDIGENOUS REMEDIAL PLANTS TO TREAT INFERTILITY DISORDERS

A total of nine species belonging to eight families were documented for indigenous therapeutic plants to treat infertility disorders (Table 2.3). The survey indicated seven species with a single family and Amaranthaceae family with two species. The juice of leaves with bioingredients was used in the preparation of natural medicine. The diversity of contemporary medicine schemes to improve contemporary revolutionary medicine is intended in ethnic vicinity, with augmentation of pharmaceutical and pharmacognostic formulas.

2.3.4   INDIGENOUS REMEDIAL PLANTS TO TREAT BONE FRACTURE

Seven species each belonging to 7 families were observed on indigenous therapeutic plants to treat bone fracture (Table 2.4), and the information on these therapeutic plants were documented. The current evidence on untamed curative plants was also recorded, except for thoroughness on their significance to train the auxiliary age group.

2.3.5   INDIGENOUS REMEDIAL PLANTS TO TREAT NEPHROLOGY DISORDERS

Total of 14 species representing 10 families were documented on natural therapeutic plants to treat kidney disorders (Table 2.5). If the wisdom on these medicinal plants is not transferred to the next generation, then such plants may vanish from our culture (Figs. 2.1 and 2.2).

2.4   SUMMARY

Indigenous remedial plants and their wisdom have been deep rooted in Indian aboriginal cultures. The wisdom of aboriginal curative plants is still being practiced. This chapter includes details on the ethnic acquaintance of 72 indigenous medicinal plants to cure skin, jaundice, bone fracture, infertility, and kidney disorders by pastoral people of Achampet forests in Telangana state of India. Out of 72 plants, the rural wisdom of original

TABLE 2.3 Indigenous Medicinal Plants for Treating Infertility Disorders.

Botanical name	Family	Habitat	Local name	Part of plant used
Achyranthus aspera	*Amaranthaceae*	Herb	Uttareni (Telugu) Utharen (Hindi)	Root
Asperagus racemosus	*Liliaceae*	Herb	Shathamuli (Telugu & Hindi)	Leaves
Barleria prionitis	*Acanthaceae*	Shrub	Bangaru kanakambralu (Telugu), Kuranta (Hindi)	Leaves
Bryonopsis laciniosa	*Cucurbitaceae*	Herb	Buddmagummadi (Telugu), Shivalingi (Hindi)	Seeds
Celosia argentia	*Amaranthaceae*	Shrub	Gunugu (Telugu), Survali (Hindi)	Leaves
Ficus religiosa	*Moraceae*	Tree	Raavi (Telugu), Peepal (Hindi)	Bark
Pergularia daemia	*Asclepidaceae*	Climber	Dustapachettu (Telugu), Uthraan (Hindi)	Leaves
Solanum xanthocarpum	*Solanaceae*	Herb	Advivnkaya (Telugu), Kantakari (Hindi)	Leaves
Sygigium cumini	*Mirtaceae*	Tree	Allaneredu (Telugu), Kalajamun (Hindi)	Seeds

TABLE 2.4 Indigenous Medicinal Plants for Treating Bone Fracture.

Botanical name	Family	Habitat	Local name	Part of plant used
Acacia arabica	Mimosaceae	Tree	Thumma (Telugu), Bhabul (Hindi)	Fruit pulp
Caesalpinia bondu	Caesalpiniaceae	Shrub	Gachakaya (Telugu), Gajga (Hindi)	Seeds
Cissus quadrangularis	Vitaceae	Herb	Nalleru (Telugu), Veldt grape (Hindi)	Aerial part
Dodonaea viscosa	Sapindaceae	Shrub	Pulivavili (Telugu), Samnata (Hindi)	Leaves
Gmelina arborea	Verbenaceae	Tree	Peddagudutekku (Telugu), Gamhar (Hindi)	Bark
Peristrophe bicalyculata	Acanthaceae	Herb	Chibira chettu (Telugu), Kaka janga (Hindi)	Leaves
Phyllanthus fraternus	Euphorbiaceae	Herb	Neelausiri (Telugu), Bhuinavalah (Hindi)	Aerial part

TABLE 2.5 Indigenous Medicinal Plants for Treating Nephrology Disorders.

Botanical name	Family	Habitat	Local name	Part of plant used
Abutilon indicum (Linn.) Sweet.	Malvaceae	Climber	Thuthura benda (Telugu), Athibala (Hindi)	Root tips
Aegle marmelos L.	Rutaceae	Tree	Maaredu (Telugu), Beel patr (Hindi)	Fresh bark peel
Amaranthus spinosa L.	Amaranthaceae	Herb	Nalleru (Telugu), Adak dathura (Hindi)	Arial part
Argemone mexicana L.	Papaveraceae	Shrub	Zeeripothu allamu (Telugu), Bharbandh (Hindi)	Roots
Cassia fistula L.	Caesalpinioideae	Herb	Nela Thangedu (Telugu), Amalthas (Hindi)	Bark peel
Celosia argentea L.	Amaranthaceae	Herb	Gunugu (Telugu)	Leaves
Cocculus hirsutus (L.)	Menispermaceae	Herb	Cheepuru theega (Telugu), Bajar bele (Hindi)	Un-ripened fruit
Phyllanthus emblica L.	Euphorbiaceae	Shrub	Usiri (Telugu), Amla (Hindi)	Leaves, fruit
Phyllanthus fraternus Webster	Euphorbiaceae	Herb	Neela Usiri (Telugu), Avla (Hindi)	Young leaves, green stem
Ricinus communis	Euphorbiaceae	Shrub	Aamudmu (Telugu), Irandi (Hindi)	Roots, stem bark
Sesamum indicum L.	Pedaliaceae	Herb	Nuvvulu (Telugu), Thil (Hindi)	Roots, stem bark
Tinospora cordifolia (Willd. L)	Menispermeaceae	Climber	Thippa theega (Telugu), Guloye (Hindi)	Leaves, seeds

remedial medicinal plants consisted of 28 species to treat skin diseases, 21 species to treat jaundice, 9 species to treat infertility, and 14 species to treat kidney disorders.

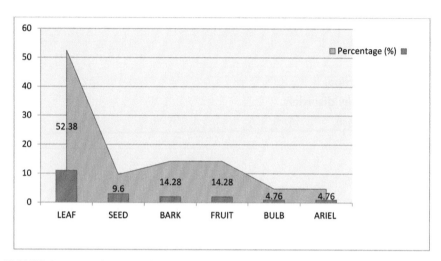

FIGURE 2.1 Fraction allocation of expansion forms of distribution via parts used.

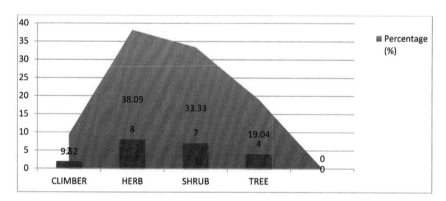

FIGURE 2.2 Fraction allocation of expansion forms of distribution via habitat.

ACKNOWLEDGMENT

Authors are thankful to rural and indigenous people of Nagarkurnool district of Telangana state for their involvement in disclosing the clandestine information.

KEYWORDS

- **bone fracture**
- **indigenous medicinal plants**
- **infertility**
- **jaundice**
- **Nagarkurnool**
- **nephrology disorder**
- **skin disease**

REFERENCES

1. Gamble, J. S. Flora of Presidency of Madras, 5th ed.; Adlard & Son Ltd.: London; 1928; Vol. I, II and III.
2. Pullaiah, T. Flora of Telangana, 1st ed.; Scientific Publishers: Jodhpur, India; 2015; Vol. I, II and III.
3. Pullaiah, T.; Chennaia, H. E. Flora of Andhra Pradesh, 1st ed.; Scientific Publishers: Jodhpur, India; 1997; Vol. I.
4. Pullaiah, T.; Moulali, D. A. Flora of Andhra Pradesh, 1st ed.; Scientific Publishers: Jodhpur, India; 1997; Vol. II.
5. Shivakumar Singh, P. The Forest Flowers and their Medicinal Properties. Vansangyan **2015**, *3* (4), 7–13.
6. Wahab, M. A.; Yousaf, M.; Hossain, M. E. Some Indigenous Medicinal Knowledge for Treating Jaundice in Chittagong Hill Tracts Bangladesh. Hamdard Medicus **2004**, *4*, 55–58.

PART II

Ethnobotany and Phytochemistry of Medicinal Plants

CHAPTER 3

PHARMACOLOGICAL PROPERTIES OF MULBERRY (*MORUS ALBA*)

NAVNEET KISHORE and DHIRAJ KUMAR

ABSTRACT

The mulberry is a famous food source of silkworms and extensively culti-vated in China and India for the production of natural silk. Mulberry plant has been widely used since 659 A.D. in traditional Chinese medicine (TCM) and well documented in the Chinese Pharmacopoeia. Various parts of the mulberry plant have resulted in various types of active metabolites during isolation including several bioactive phenolic compounds with isoprenoid substitution. The plant *Morus alba* L. has a diverse range of pharmacological properties and is used to cure many diseases in folk medicine. In many countries, it is consumed conventionally as a feed in mixed forage diets for ruminants; people are also using this for making delicious curry due to its nutritional contents. Its sole use in current medication is for the preparation of syrup to impart flavor and color in any other medicines. Mulberry fruit juice is documented as the endorsed medicine of the British Pharmaco-poeia. The present chapter focuses on the ethnopharmacological relevance, medicinal aspects, economical value, and importance of mulberry plant at industrial scale.

3.1 INTRODUCTION

Mulberry is an angiosperm plant belonging to the Moraceae family. This group of plants is commonly named as mulberries, which are either wild or cultivated in many temperate regions of the world.[31] These plants are grown for their medicinal and edible use in daily life. The most common species belonging to this group is *Morus alba* L. (white mulberry)[8], *Morus australis* (Chinese mulberry), *Morus mesozygia* (African mulberry), *Morus macroura*) long mulberry, *Morus nigra* (black mulberry), and *Morus rubra* (red mulberry).[2] White mulberry is scientifically known as *Morus Alba L., which has been used frequently in the* traditional Chinese medicine (TCM) since a long time ago (2000 years), and it has been used as traditional medicine to treat various ailments. The antibacterial activity of root bark of this plant against food poisoning bacteria has been reported in traditional medicine.

An alcoholic extract of the mulberry leaves has shown antioxidant, antihyperglycemic, and antiglycation effects in chronic diabetic rats, thus suggesting use of mulberry leaves as a food supplement for diabetic patients. The freeze-dried powder extract from fruits has displayed hypolipidemic and antioxidant properties[30], and *in vivo* and *in vitro* neuroprotective effects.[18] It also acts as good antistress agent when used as methanol extract.[22] The extract from leaves helps restore the vascular reactivity of diabetic rats[23] and has been used for potent antisnake venom activity against the specific *Daboia russelii* and vipera venom.[3]

This chapter focuses on the ethnopharmacological relevance, medicinal aspects, economical value, and importance of mulberry plant at industrial scale.

3.2   BRIEF DESCRIPTION OF WHITE MULBERRY (*MORUS ALBA* L.)

Morus alba L. (white mulberry) usually originated from Asia and is used for sericulture in India, China, Vietnam, and Korea. It is also known as the "folk sacred fruit". The mulberry trees have been extensively scattered throughout the world representing different species, which are native to other countries, such as America, Arabia, China, India, Japan, North Africa, South Europe, and Taiwan etc. It is a medium sized tree about 30 m in height and monoecious or deciduous tree with large stems bearing bark. This species is indigenous to China and is also found in hilly areas of Himalayas. This species has been used in China as a remedy to various diseases.

3.3   MULBERRY: IMPORTANCE OF RESEARCH EXPLORATION

Mulberry is grown under various climatic conditions and is distributed ranging from tropical to temperate regions throughout the globe. It is one of the dominant species among 24 known species. India holds the second largest area of mulberry (>280,000 ha) after China having about 626,000 ha. This plant has vast usage of leaves and fruits for its medicinal and edible potential. They can be also used for making tea leaves, and are an efficient diet for silkworms in the production of silk at a commercial level.[12] The white mulberry is widely cultivated in its native states and is the primary food source for silk moths.

The fruit juice is a good health beverage in Asian countries. The fresh fruits of this plant are eaten directly, whereas dried fruits are processed into jam, wine and fruit juice due to its pleasant color, and low-calorie content with high nutrient value as well as delightful taste. Whole plant parts, such as leaves, roots, root bark, twinges, and fruits, have medicinal properties.[32] The biological potential is attributed to the presence of countless secondary metabolites. The main bioactive ingredients identified from *M. alba* are alkaloids, Diels–Alder type adducts, flavonoids, polysaccharides, and stilbenes.[11,36] Based on the diverse range of usage in the medicine and commercial level, it should be a significant part of today's research and development.

The better use of mulberry plant onset of the available information will signpost to researchers for further investigations to prove properties of *M. alba*. This signifies focus on the data compilation on this plant. Moreover, several secondary metabolites have been identified from mulberry plant, which are on the increase every day for research going on this plant. Consequently, the pharmacological properties of specific isolated compounds and their mechanism of action for a particular health benefit are still warranted. All these much needed research and development can be established through further exploration of this plant.

3.4 PHARMACOLOGICAL PROPERTIES OF WHITE MULBERRY

Almost all parts from the white mulberry, such as leaves, branches, fruits, and root bark, have been extensively used in the Chinese drug prescription for the treatment of several ailments.[29] Leaves are used to treat arthritis and hypertension, and the root bark is well known as a vital drug to treat heart diseases, inflammation, cancer, hepatitis, diabetes, and cough in addition to prepare the new generation medicines and formulation of syrup to impart taste. The bark and roots of mulberry are significant ordinary medicine to treat cough, dyspnea, inflammatory disorders, heart diseases, hepatitis, lung-fire, cancer, hypertension, diuresis, and diabetes. "*Sohaku-hi*" is a famous Chinese medicine extracted from the mulberry root to decrease plasma sugar level in mice.[34] Leaves are also useful to keep the skin smooth and healthy, and to prevent inflammation, throat infection, diaphoretic, irritation, and emollient. Mulberry is an excellent source of anthocyanins and can possibly be used for the production of anthocyanin on industrial level.[19]

The use of root bark mulberry species by human beings has been mentioned in ancient literature. The antidiabetic agents (piperidine alkaloid and glycoproteins) have been isolated from the leaves and roots. Another research study exhibited antioxidant potential of some phenolic compounds in mulberry species. The *Morus alba* extract exhibited significant inhibition of tyrosinase and hyperpigmentation issues.[1] The flavonoid-rich fraction from 70% alcoholic extract of root bark displayed good hypoglycemic potential.[25] The root bark of *Morus* plant and isolated phytocompounds showed good pancreatic lipase inhibitory activity.[9] The development of antiobesity drug induced by diet has attributed to pancreatic lipase (PL), which is known as one of the harmless targets. Therefore, PL performs a vital function to prevent obesity. The ethyl acetate fraction displayed in vivo antidepressant effects and led to the isolation of phytochemicals with good biological activity. Active metabolites isolated from mulberry plant have shown significant action on cell death in HT22 cells induced by glutamate to prove neuroprotective potential[7] and hepatoprotective action in HepG2 cells during tert-butyl hydroperoxide (t-BHP)-induced oxidative stress.[14,15] The methanol extract exhibited significant cytotoxic effect against Hep3B, MCF-7 (Michigan Cancer Foundation-7), and HeLa cell lines.[7]

3.5   WHITE MULBERRY: MEDICINAL VALUE OF FRUITS

The juice from mulberry fruits is dark violet or purple liquid with a sweet taste. This juice is used as syrup in various folk medicines as relief from sore throat, remunerations to liver and kidney, blood purifier, nourishing the skin, and treatment of tinnitus, urinary incontinence, constipation, depression, fever, dizziness, calming the nerves, balancing internal secretions, metabolism of alcohol, and to enhance immunity. Moreover, the mulberry fruits can be utilized to prepare wine and resin substitutes. Different varieties of mulberry fruits traditionally found in Asia have shown various biochemical activities, namely, inhibition of gathering of hepatic triglyceride[35], antitumor properties[6], antioxidant activities[4], hypolipidemic possession[21], antidiabetic activity[27], atherosclerosis prevention[20,26], and antiobesity.[24] It is a very good source of anthocyanins that can possibly be used for the production of anthocyanin on industrial level.

3.6 BIOACTIVE INGREDIENTS ISOLATED FROM MULBERRY

The chemical composition among different mulberry species varies significantly.[13] Literature survey revealed that the *Morus* plant contains natural isoprenoid substituted phenolic compounds in significant amounts, and these isoprenes have effective potential for pharmacological evaluation. Glucosidase is a drug isolated from the roots of this plant to treat high blood pressure. The root decoction also helps in the blood adherence and in the digestive ailments by killing the worms.

Various active metabolites including prenylated flavonoids and benzofurans have been reported earlier from this plant with pharmacological actions, such as inhibition of lipooxygenase-1 (LOX-1) and nuclear factor kappa B (NF-κB), antioxidant, cytotoxicity, hepatoprotection, invasion and migration of cancer cells, and excellent glycosidase inhibitory action due to several alkaloids.[7] The extract of mulberry leaves contains rutin, isoquercetin, several derivatives of kaempferol, and quercetin glycosides that exhibit strong antioxidant action.[5] Albanol-A compound isolated from root bark indicated significant potential to treat leukemia.[16] Also steppogenin-4'-*O*-*β*-D-glucoside, moracin-M, and mulberroside-A compounds showed positive hypoglycemic effects.[33] Mulberroside-A, a glycosylated stilbenoid, can be used to treat gout and hyperuricemia.[17,28]

The anticancer potential of morusin and other related flavonoids with prenylated substitutions were assessed with MMT (3-(4,5-dimethylthiazol2-yl)-2,5-diphenyl-tetrazolium) assay. The significant inhibition was observed by the morusin with an IC_{50} (50% inhibitory concentration) value of 0.64 μM in contrast to HeLa cancer cell lines. It was also effective against colorectal cancer cells HT-29 with an IC_{50} value of 29 μM by the inhibition of NF-kB signaling in humans. Therefore, the anticancer potential of morusin on HeLa cell lines exhibited comparable mechanistic way, which could be established through further investigation on HeLa cells.

Research studies have revealed identification of several classes of secondary metabolites, such as adducts of Diels–Alder types, benzofurans, coumarins, flavonoids with isoprenylated moieties, triterpenoids, and stilbenes. The methanol extract from air-dried powdered leaves led to the isolation of several active metabolites, such as atalantoflavone, cyclomorusin, cyclomulberrin, 8-geranylapigenin, diprenyl-trihydroxyflavone, geranyl-prenyl-tetrahydroxy-flavone, kaempferol, kuwanon-S, morusin, sanggenon-J, and sanggenon-K.[7] Metabolites were isolated showing pancreatic lipase inhibitory activity in another investigation. These compounds were[9]:

cudraflavone-C	moracin-N	sanggenofuran-A
cyclomorusin	morusin	sanggenol-A
isobavachalcone	mulberrofuran-A	sanggenol-L
kenusanone-A	mulberrofuran-C	sanggenol-O
kuwanon-T	mulberrofuran-D	sanggenol-Q
kuwanon-U	mulberrofuran-D2	sanggenon-F
moracin-B	mulberrofuran-G	sanggenon-N
moracin-C	mulberrofuran-H	sepicanin-A
moracin-E	mulberrofuran-W	

3.7 FUTURE PERSPECTIVES

Mulberry plant is eminent not only for its medicinal value, but also as efficient nutrient food. The herbal products of this plant have added significant devotion in the last decade. The countless bioactive metabolites have been isolated from this plant, which displayed significant anti-inflammatory, immunomodulatory, antidiabetic, antiobesity, antioxidant, hepatoprotective, and renoprotective activities. The fruits and leaves of this plant are good feed in mixed forage for ruminants. Moreover, this plant is the most efficient host of silk production at industrial scale because the leaves are favorite food for silk moths (*Bombyx mori*).

Despite diverse range of ethnopharmacological, pharmacological, phytochemical, and other significant research studies on *M. alba*, the knowledge on the biological potential of metabolites and their mode of action is still lacking. However, the previous reports indicate a number of secondary metabolites from this plant and there may be many more unknown biocompounds that may be responsible for the biological action. Hence, the further research needed for the identification of unknown metabolites and the structure–activity relationship of their respective biological function. Therefore, the identified entities should be considered for clinical trials to explore the mechanism of action. In recent years, many polysaccharide derivatives[10] have been identified from the *M. alba* and further research study should establish their unambiguous structures and biological functions.

3.8 SUMMARY

The present chapter compiles the ethnopharmacological, pharmacological, phytochemical profiles, and recent updates on *M. alba*. More than 200 metabolites have been identified from this plant and yet the number is increasing due to day by day research efforts. In this chapter, authors have summarized the whole story of this plant and have included identified biocompounds with prominent activity. All data were collected from the various search engines like SciFinder, Google Scholar, and PubMed by entering different keywords. More than 100 research articles are available and well documented in the literature. A very recent report has shown that the polysaccharides obtained from *M. alba* significantly lowered the level of blood sugar. However, further investigation is required to show the relationship by knowing the mechanism of action. This chapter provides the platform for researchers and practitioners who are working on *M. alba* plant.

ACKNOWLEDGMENTS

The authors sincerely acknowledge the University of Delhi, India and Jiwaji University, India for the research support. University of Grant Commission (UGC), New Delhi is sincerely acknowledged for the award of Dr D. S. Kothari postdoctoral fellowship for financial support.

KEYWORDS

- **economic value**
- **ethnopharmacology**
- ***Morus alba* L.**
- **mulberry plant**

REFERENCES

1. Baurin, N.; Arnoult, E.; Scior, T.; Do, Q. T.; Bernard, P. Preliminary Screening of Some Tropical Plants for Anti-Tyrosinase Activity. *J. Ethnopharmacol.* **2002,** *82*, 155–158.

2. Butt, M. S.; Nazir, A.; Sultan, M. T.; Schroen, K. *Morus alba* L. Nature's Functional Tonic. *Trend. Food Sci. Technol.* **2008,** *19*, 505–512.

3. Chandrashekara, K. T.; Nagaraju, S.; Nandini, U. S.; Basavaiah, K. K. Neutralization of Local and Systemic Toxicity of *Daboia russelii* Venom by *Morus alba* Plant Leaf Extract. *Phytother. Res.* **2009,** *23*, 1082–1087.

4. Chang, L. W.; Juang, L. J.; Wang, B. S.; Wang, M. Y.; Tai, H. M.; Hung, W. J.; Chen, Y. J.; Huang, M. H. Antioxidant and Antityrosinase Activity of Mulberry (*Morus alba* L.) Twigs and Root Bark. *Food Chem. Toxicol.* **2011,** *49*, 785–790.

5. Chao, P.-Y.; Lin, K.-H.; Chiu, C.-C.; Yang, Y.-Y.; Huang, M.-Y.; Yang, C.-M. Inhibitive Effects of Mulberry Leaf-Related Extracts on Cell Adhesion and Inflammatory Response in Human Aortic Endothelial Cells. *Evid.-Based Complement. Altern. Med.* **2013,** *2013*, 1–14.

6. Chen, M. G.; Jin, P. H.; Huang, L. X. Energy Analysis of Mulberry-Silkworm Ecosystem in China. *Chin. J. Appl. Ecol.* **2006,** *17*, 233–236.

7. Dat, N. T.; Binh, P. T. X.; Quynh, L. T. P.; Minh, C. V.; Huong, H. T.; Lee, J. J. Cytotoxic Prenylated Flavonoids from *Morus alba. Fitoterapia* **2010,** *81*, 1224–1227.

8. Ercisli, S.; Orhan, E. Chemical Composition of White (*Morus Alba*), Red (*Morus rubra*), and Black (*Morus nigra*) Mulberry Fruits. *Food Chem.* **2007,** *103*, 1380–1384.

9. Ha, M. T.; Tran, M. H.; Ah, K. J.; Jo, K.-J.; Kim, J.; Kim, W. D.; Cheon, W. J.; Woo, M. H.; Ryu, S. H.; Min, B. S. Potential Pancreatic Lipase Inhibitory Activity of Phenolic Constituents from the Root Bark of *Morus alba* L. *Bioorg. Med. Chem. Lett.* **2016,** *26*, 2788–2794.

10. He, X.; Fang, J.; Ruan, Y.; Wang, X.; Sun, Y.; Wu, N.; Zhao, Z.; Chang, Y.; Ning, N.; Guo, H.; Huang, L. Structures, Bioactivities and Future Prospective of Polysaccharides from *Morus alba* (White Mulberry): A Review. *Food Chem.* **2018,** *245*, 899–910.

11. Huang, L. X.; Zhou, Y. B.; Meng, L. W.; Wu, D.; He, Y. Comparison of Different CCD Detectors and Chemometrics for Predicting Total Anthocyanin Content and Antioxidant Activity of Mulberry Fruit Using Visible and Near Infrared Hyperspectral Imaging Technique. *Food Chem.* **2017,** *224*, 1–10.

12. Jelled, A.; Hassine, R. B.; Thouri, A.; Flamini, G.; Chahdoura, H.; Arem, A. E.; Lamine, J. B.; Kacem, A.; Haouas, Z.; Cheikh, H. B. Immature Mulberry Fruits Richness of Promising Constituents in Contrast with Mature Ones: A Comparative Study Among Three Tunisian Species. *Ind. Crop. Prod.* **2017,** *95*, 434443.

13. Jiang, D.-Q.; Guo, Y.; Xu, D.-H.; Huang, Y.-S.; Yuan, K.; Lv, Z.-Q. Antioxidant and Anti-Fatigue Effects of Anthocyanins of Mulberry Juice Purification (MJP) and Mulberry Marc Purification (MMP) from Different Varieties Mulberry Fruit in China. *Food Chem. Toxicol.* **2013,** *59*, 1–7.

14. Jung, J.-W.; Ko, W.-M.; Park, J.-H.; Seo, K.-H.; Oh, E.-J.; Lee, D.-Y.; Lee, D.-S.; Kim, Y.-C.; Lim, D.-W.; Han, D.; Baek, N. I. Isoprenylated Flavonoids from the Root Bark of *Morus alba* and their Hepatoprotective and Neuroprotective Activity. *Arch. Pharmacol. Res.* **2015,** *38*, 2066–2075.

15. Jung, J.-W.; Park, J.-H.; Lee, Y.-G.; Seo, K.-H.; Oh, E.-J.; Lee, D.-Y.; Lim, D.-W.; Han, D.; Baek, N.-I. Three New Isoprenylated Flavonoids from the Root Bark of *Morus alba. Molecules* **2016,** *21*, 1112.

16. Kikuchi, T.; Nihei, M.; Nagai, H.; Fukushi, H.; Tabata, K.; Suzuki, T.; Akihisa, T. Albanol a from the Root Bark of *Morus alba* L. Induces Apoptotic Cell death in HL60 Human Leukemia Cell Line. *Chem. Pharmaceut. Bull.* **2010,** *58*, 568571.

17. Kim, H. G.; Ju, M. S.; Shim, J. S.; Kim, M. C.; Lee, S. H.; Huh, Y.; Kim, S. Y.; Oh, M. S. Mulberry Fruit Protects Dopaminergic Neurons in Toxin-Induced Parkinson's Disease Models. *Br. J. Nutr.* **2010**, *104*, 8–16.

18. Kim, J. K.; Kim, M.; Cho, S. G.; Kim, M. K.; Kim, S. W.; Lim, Y. H. Biotransformation of Mulberroside a from *Morus Alba* Results in Enhancement of Tyrosinase Inhibition. *J. Ind. Microbiol. Biotechnol.* **2010**, *37*, 631–637.

19. Kumar, V. R.; Chauhan, S. Mulberry: Life Enhancer. *J. Med. Plant Res.* **2008**, *2*, 271–278.

20. Liu, L. K.; Lee, H. J.; Shih, Y. W.; Chyau, C. C.; Wang, C. J. Mulberry Anthocyanin Extracts Inhibit LDL Oxidation and Macrophage Derived Foam Cell Formation Induced by Oxidative LDL. *J. Food Sci.* **2008**, *73*, H113–H121.

21. Liu, L. K.; Chou, F. P.; Chen, Y. C.; Chyau, C. C.; Ho, H. H.; Wang, C. J. Effects of Mulberry (*Morus alba* L.) Extracts on Lipid Homeostasis *In Vitro* and *In Vivo*. *J. Agric. Food Chem.* **2009**, *57*, 7605–7611.

22. Nade, V. S.; Kawale, L. A.; Naik, R. A.; Yadav, A. V. Adaptogenic Effect of *Morus Alba* on Chronic Footshock-Induced Stress in Rats. *Indian J. Pharmacol.* **2009**, *41*, 246–251.

23. Naowaboot, J.; Pannangpetch, P.; Kukongviriyapan, V.; Kukongviriyapan, U.; Nakmareong, S.; Itharat, A. Mulberry Leaf Extract Restores Arterial Pressure in Streptozotocin-Induced Chronic Diabetic Rats. *Nutr. Res.* **2009**, *29*, 602–608.

24. Peng, C. H.; Liu, L. K.; Chuang, C. M.; Chyau, C. C.; Huang, C. N.; Wang, C. J. Mulberry Water Extracts Possess an Anti-Obesity Effect and Ability to Inhibit Hepatic Lipogenesis and Promote Lipolysis. *J. Agric. Food Chem.* **2011**, *59*, 2663–2671.

25. Singab, A. N. B.; El-Beshbishy, H. A.; Yonekawa, M.; Nomura, T.; Fukai, T. Hypoglycemic Effect of Egyptian *Morus Alba* Root Bark Extract: Effect on Diabetes and Lipid Peroxidation of Streptozotocin-Induced Diabetic Rats. *J. Ethnopharmacol.* **2005**, *100*, 333–338.

26. Tang, C.-C.; Huang, H.-P.; Lee, Y.-J.; Tang, Y.-H.; Wang, C.-J. Hepatoprotective Effect of Mulberry Water Extracts on Ethanol-Induced Liver Injury Via Anti-Inflammation and Inhibition of Lipogenesis in C57BL/6J Mice. *Food Chem. Toxicol.* **2013**, *62*, 786–796.

27. Thirugnanasambandham, K.; Sivakumar, V.; Maran, J. P. Microwave Assisted Extraction of Polysaccharides from Mulberry Leaves. *Int. J. Biol. Macromol.* **2015**, *72*, 1–5.

28. Wang, C.-P.; Wang, Y.; Wang, X.; Zhang, X.; Ye, J.-F.; Hu, L. S.; Kong, L. D. Mulberroside A. Possesses Potent Uricosuric and Nephroprotective Effects in Hyperuricemic Mice. *Planta Med.* **2011**, *77*, 786–796.

29. Wang, X.; Deng, Q. F.; Chen, G. H.; Zhou, X. Characterization and Activity Effect on ADH of Polysaccharides from *Mori Fructus*. *China J. Chin. Mater. Med.* **2017**, *42*, 2329–2333.

30. Yang, X.; Yang, L.; Zheng, H. Hypolipidemic and Antioxidant Effects of Mulberry (*Morus Alba*) Fruit in Hyperlipidaemia Rats. *Food Chem. Toxicol.* **2010**, *48*, 2374–2379.

31. Yuan, Q.; Zhao, L. The Mulberry (*Morus Alba* L.) Fruit: A Review of Characteristic Components and Health Benefits. *J. Agric. Food Chem.* **2017**, *65*, 1038310394.

32. Zhang, D. Y.; Wan, Y.; Xu, J. Y.; Wu, G. H.; Li, L.; Yao, X. H. Ultrasound Extraction of Polysaccharides from Mulberry Leaves and Their Effect on Enhancing Antioxidant Activity. *Carbohydr. Polym.* **2016**, *137*, 473–479.

33. Zhang, M.; Chen, M.; Zhang, H.-Q.; Sun, S.; Xia, B.; Wu, F.-H. In Vivo Hypoglycemic Effects of Phenolics from the Root Bark of *Morus alba*. *Fitoterapia* **2009**, *80*, 475–477.

34. Zhang, Y.; Ren, C.; Lu, G.; Cui, W.; Mu, Z.; Gao, H.; Wang, Y. Purification, Characterization and Anti-Diabetic Activity of a Polysaccharide from Mulberry Leaf. *Regul. Toxicol. Pharmacol.* **2014**, *70*, 687–695.

35. Zhang, Y.; Ren, C.; Lu, G.; Mu, Z.; Cui, W.; Gao, H.; Wang, Y. Anti-Diabetic Effect of Mulberry Leaf Polysaccharide by Inhibiting Pancreatic Islet Cell Apoptosis and Ameliorating Insulin Secretory Capacity in Diabetic Rats. *Int. Immunopharmacol.* **2014,** *22*, 248–257.

36. Zhou, M.; Chen, Q. Q.; Bi, J. F.; Wang, Y. X.; Wu, X. Y. Degradation Kinetics of Cyanidin 3-O-Glucoside and Cyanidin 3-Orutinoside During Hot Air and Vacuum Drying in Mulberry (*Morus alba* L.) Fruit: A Comparative Study Based on Solid Food System. *Food Chem.* **2017,** *229,* 574–579.

CHAPTER 4

ETHNOBOTANY, PHYTOCHEMISTRY, AND PHARMACOLOGY OF GENUS *RHODIOLA* (L.): POTENTIAL MEDICINAL APPLICATIONS

SAHIL KAPOOR, ASHWANI KUMAR BHARDWAJ, ASHISH RAMBHAU WARGHAT, KUSHAL KUMAR, AVILEKH NARYAL, and OM PRAKASH CHAURASIA

ABSTRACT

Rhodiola is a perennial plant with a fleshy caudex, crowned with leaves and a broad clasping base often reduced to membranous or semi-orbicular scales from the axils of which leafy flowering-shoots are produced. There are approximately 23 species of *Rhodiola*, which are distributed in the Himalaya region. *Rhodiola imbricata* is used to treat various epidemic diseases, traumatic injuries and burns, edema of legs, and for eliminating toxins from the body. Besides these activities, it can reduce the effects of anoxia and fatigue. It can also provide beneficial effects against various neurological disorders. This chapter reviews the morphometry, phytochemical profiling, and traditional usage of *Rhodiola* plant in the Himalaya region.

4.1 INTRODUCTION

The term "Ethnobotany" was coined in 1895 by US Botanist John William Harshberger, who defined ethnobotany as "the subject that helps in elucidating the cultural position of tribes who use plants for food and shelter, helps in suggesting new lines of manufacture and provides important information regarding the socio-ecological history of plants". The field of ethnobotany has experienced a great paradigm shift from being the raw compilation of data to an organized methodological reorientation.[9] Some of the important drugs of the modern pharmaceutical era like atropine and digoxin have come into practice only due to the existing traditional knowledge and meticulous study of the indigenous remedies involving phytomedicines. Ethnobotany usually comprises three main aspects: traditional, modern, and environmental.

Rhodiola is a perennial plant with a fleshy caudex, crowned with leaves, and a broad clasping base often reduced to membranous or semi-orbicular scales from the axils of which leafy flowering shoots are produced.[26] *Rhodiola* is generally distributed along the alpine habitats in high-altitude terrains and grows along the rocky slopes and stony crevices. Several traditional communities across the globe have been using *Rhodiola* for therapeutic purposes. For example, Norwegians extensively use the plant to treat urinary disorders, scurvy, and lung diseases.[4] The rhizome of *Rhodiola* species is rich in secondary metabolites, such as polyphenols, phenol glycosides, and organic acid that are of high therapeutic value.[25] *Rhodiola* has been found to treat various stress disorders associated with acute and chronic exposure to hypobaric hypoxia, which has been associated with various cognitive dysfunction.[4,5,11,20] Several pharmacological studies on *Rhodiola* have demonstrated that it exhibits cardioprotective, antipyretic, anti-stress, anti-inflammatory, antioxidant, anti-aging, anti-depressant, and radioprotective properties.[5,6,8,11,19]

This chapter discusses medicinal properties of *Rhodiola* that has immense therapeutic applications, provided its species are accurately identified. Thus, segregation and identification of different species of *Rhodiola* on the basis of morphometry, phytochemical profiling, and traditional usage shall help in effective utilization and development of pharmacologically valuable biomolecules that could considerably benefit mankind.

4.2 BOTANICAL CLASSIFICATION AND DISTRIBUTION OF *RHODIOLA SP.*

4.2.1 *DISTRIBUTION OF RHODIOLA SP.*

The word *Rhodiola* is derived from the Greek word "rhodon", meaning rose and referring to the rose-like smell of roots and Latin word "iola" that means diminutive. Based on GBIF records, there are about 23 species of *Rhodiola*, which are distributed in the Himalaya region (Fig. 4.1), and one new species of *Rhodiola* (i.e. *Rhodiola sedoides*) has been documented in Arunachal Pradesh of India.[22] Table 4.1 indicates morphological description of some species of *Rhodiola* indigenous to India.

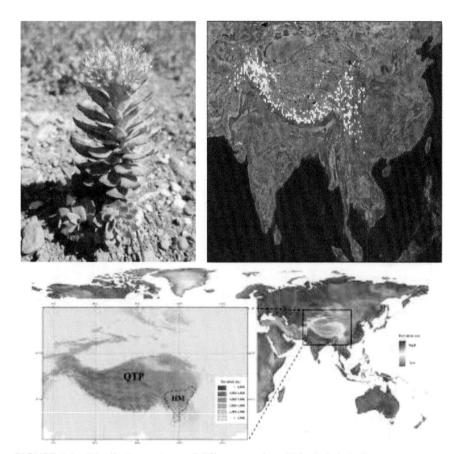

FIGURE 4.1   Distribution patterns of different species of *Rhodiola* in India.

4.3   TRADITIONAL USES OF SOME *RHODIOLA* SP.

Rhodiola imbricate is a wild edible plant, and the young shoots of *Rhodiola* are boiled and mixed with yogurt to prepare a local Ladakh delicacy "Tantur".[6] Since ancient times, the roots and rhizome of *Rhodiola imbricate* have been used by local people for various medicinal purposes.[12] In traditional herbal medicine system, it is used for the treatment of various epidemic diseases, traumatic injuries and burns, edema of legs, and for eliminating toxins from the body. Besides these properties, it is known to reduce the effects of anoxia and fatigue. In Amchi and Tibetan system of traditional medicine, roots have medicinal usage against respiratory ailments, cold, cough, and contagious diseases.[7]

TABLE 4.1 Morphological Description of Selected *Rhodiola* Species.

Botanical name	Habitat	Plant description	Flowering and fruiting	References
Rhodiola quadrifida	Alpine and Arctic regions, Stony slopes, rocks; Sikkim; 7546-13,451 ft.	It is a glabrous herbaceous perennial with stout and elongated rhizome. Stems are 15 cm long, erect, and simple. Leaves are linear, acute, and flattened. Flowers are tetra- or penta-parted. The male flower is linear-lanceolate, blunt, wide spreading, and usually purplish. Stamens are erect.	May to September	[16, 24]
Rhodiola bupleuroides	Grassy places, rock crevices on slopes; Himalayas (Kumaon, Uttarakhand), 10,000-12,000 ft.	It is a glabrous perennial herb with a massive rootstock. Stem is erect and about 22-30 cm long. Leaves are alternate, fleshy, ovate, sessile, green, and paler on the abaxial side. Inflorescence is a flat leafy cyme, which is 3-8 cm across. Buds are obovate or nearly globular. Flowers are penta-parted, and male flowers have linear sepals with oblanceolate petals and often reflexed; stamens are reddish purple and shorter than the petals. The female flowers have more linear petals; carpels are large and purple in color with short blunt and linear spreading styles.	June to September	[14, 35]
Rhodiola chrysanthemifolia	Grasslands, rocks, rock crevices; Himalayas	It is a perennial herb with thick branched rootstock, and is 6-7 mm in diameter.	August to October	[15]

TABLE 4.1 *(Continued)*

Botanical name	Habitat	Plant description	Flowering and fruiting	References
	(Nepal), 10,500 –13,780 ft.	Stem is erect and about 10–25 cm long.		
		Leaves are scale like, ovate with obtuse apex.		
		Inflorescence is a compact corymbi form.		
		Flowers are bisexual, unequally pentamerous; sepals are linear, petals are oblong ovate, 7–9 × 1.5–3 mm with entire margin, stamens are 10 in number.		
Rhodiola coccinea	Alpine regions, stony soils, and rocks; NW India, Kashmir; 8530-16,076 ft.	It is a perennial herb with long and thick branched roots.	June to August	[33]
		Stem is cylindrical, 1.5-2 mm thick and about 8-38 cm long.		
		Leaves are alternate, fleshy, sessile, entire, sub-obtuse, 5-6 mm long and 1-1.5 mm broad.		
		Inflorescence is a compact corymbiform, which is 0.8-1 cm in diameter.		
		Flowers are dioecious, usually pentamerous and longer than pedicels; sepals are oblong and sub-obtuse; petals are oblong-ovate, reddish, and constricted at the apex. Stamens are 10 in number; anthers are yellow and rounded. Seeds are 1-1.5 mm long, and are brown and oblong.		
Rhodiola cremulata	Grassland slopes, mountain slopes, rocky places, and rock crevices; Sikkim and Lahul; 9186-18,372 ft.	It is a perennial herb up to 20 cm tall with a thick sub-cylindrical rhizome.	June to August	[15]
		Stem is fastigiated and glabrous.		
		Leaves are alternate, more or less indistinctly petiolate.		

TABLE 4.1 *(Continued)*

Botanical name	Habitat	Plant description	Flowering and fruiting	References
Rhodiola cretinii	Rock crevices; NE India (Assam), Sikkim; 12,140–13,451 ft.	Inflorescence is dense or compact corymbose fascicles surrounding by the upper leaves and bracts. It is a perennial herb with thick branched roots. Stem is sub-erect and crowned with leaves. Leaves are narrowly elliptic, linear and obovate with 7-10 × 1.5-2.5 mm recurrent base. Inflorescence is a dense corymbiform. Flowers are unisexual and unequally pentamerous. Sepals are linear, 3-4.5 mm with obtuse apex. Petals are green to yellowish white, linear-oblanceolate, narrowly elliptic, 3.5–6 × 1–1.5 mm with obtuse apex. Stamens are 10 in number. Carpels are erect, narrowly ellipsoid, 5–7 mm.	June to August	[16]
Rhodiola fastigiata	Rocky slopes; Kashmir and Sikkim; 11,482–17,716 ft.	It is a glabrous perennial herb with thick branched caudex. Stem is smooth, erect, and about 8-15 cm long. Leaves are alternate, linear-oblong to lanceolate. Inflorescence is a compact corymbiform. Flowers are tetra- or penta-parted; male flower has sepals that are linear to long triangular; petals are blunt and broad lanceolate; stamens are spreading and twice the sepals. Female flower has oblong sepals; petals are linear, blunt, strap-shaped, emarginated and reflexed; carpels are very erect, slightly longer than the petals.	June to August	[15]

TABLE 4.1 (*Continued*)

Botanical name	Habitat	Plant description	Flowering and fruiting	References
Rhodiola heterodonta	Slopes, sides of ravines; Kashmir, Ladakh (Khardungla, Changla.	It is a glabrous perennial herb with thick, elongated root stock, and occasionally branched, similar to *Rhodiolarosea*.	June to September	[16,34]
		Stem is smooth and erect.		
	and Penzila) and Kumaon, 9186–18,500 ft	Leaves are alternate, triangular to ovate, form a clasping base, coarsely toothed, fleshy, green or glaucous, and is loosely disposed on the stem.		
		Inflorescence is terminal and dense.		
		Flowers are pentamerous on very short pedicels. Male flower has sepals that are linear, blunt and greenish; petals are blunt, yellowish or reddish; stamens are widely spread and slightly exceed the petals, and filaments are streaked red. Female flower has sepals and petals are of the same length, linear, erect, blunt, green or purplish; carpels are large, erect, green, and style is very short in size.		
Rhodiola himalensis	Forests, scrub, slopes; Sikkim; 12,000–17,000 ft.	It is a glabrous herbaceous perennial with thick, elongated root stack armed with the bases of the old stem, and crowned with conspicuous scale-leaves.	May to August	[15]
		Stem is erect, round, usually reddish, 15–30 cm long, leafy throughout and mostly rough with many transparent bead-like glands.		
		Leaves are alternate, sessile, fleshy, lanceolate to oblanceolate, oblong–oblanceolate, acute to apiculate, toothed near apex or entire, dark green and paler below.		

TABLE 4.1 *(Continued)*

Botanical name	Habitat	Plant description	Flowering and fruiting	References
		Flowers are dark purplish with cylindrical pedicels. Male flower has calyx lobes tapering from a broad base; petals are oblong lanceolate, blunt, twice the calyx with ascending red filaments and anthers are deep red. Female flower has erect stout carpels with a very short diverging style.		
Rhodiola hobsonii	Rock crevices; Sikkim; 8530–13,451 ft.	It is a perennial herb with thick, elongated rootstock. Stem is erect, round and 6–13 cm long.	August to September	[30]
		Leaves are alternate, sessile, and narrowly oblanceolate to oblong with entire margin and acute apex.		
		Flowers are unequally pentamerous; sepals are triangular to ovate-lanceolate; petals are ovate with connate base; stamens are 10 in number and anteseplous one; carpels are 5.5 mm long and styles are 1 mm long; seeds are ovoid and 0.8 mm long.		
Rhodiola imbricata	Rocky slopes, wet places; Ladakh (Khardungla, Changla and Penzila) and Kumaon; 12,000– 18,380 ft.	It is a dioecious perennial herb with sub-cylindrical rhizome, which is sparsely branched and 2-2.5 cm long. Stem is glabrous, 10-20 cm long.	June to August	[10]
		Leaves are densely arranged, oblanceolate to narrowly elliptic, 2-3 cm × 37 mm, sessile, glabrous with acute tip, and round base.		
		Petals are angular-oblanceolate; stamens are distinctly longer than the petals; filaments are 5-8 mm; anthers are dark purplish red; carpels are 3-5 mm with 8-10 ovules.		

TABLE 4.1 (Continued)

Botanical name	Habitat	Plant description	Flowering and fruiting	References
Rhodiola pamiroalaica	Valley slopes, rock crevices; Lahul valley; 7874-9186 ft.	It is a perennial herb with robust thick roots.	June to August	[5]
		Leaves are alternate, linear-lanceolate, 7-15 cm × 1-2 mm, sessile, glabrous with broad base and entire margin.		
		Inflorescence is a compact corymbiformpaniculate 1 × 1-2 cm in diameter.		
		Flowers are unequally pentamerous in male plants; sepals are greenish yellow, linear to lanceolate, 2-3 mm with sub-obtuse apex; petals are lanceolate to linear, 4-5 mm with sub-obtuse apex; stamens are 10-12 in number and are shorter than the petals.		
Rhodiola praini	Rocks in broad-leaved forests in valleys, rock crevices; NE India (Assam), Sikkim; 7217-14,107 ft.	It is a perennial herb with thick, elongated root stack.	August to September	[30]
		Stem is erect, round and about 8 cm long.		
		Leaves are verticillate, ovate, glabrous with entire margin and base, which is abruptly narrowed to long attenuate.		
		Simple or branched corymbiform inflorescence, which is 1-4 cm in diameter.		
		Flowers are bisexual and unequally pentamerous with 1.5–2.5 mm pedicel; sepals are narrowly triangular ovate with acute apex and obtuse tip; petals are white, pink or red, oblong- ovate with arose margin; stamens are 10 in number and are shorter than petals, and carpels are lancelate to ovoid.		

TABLE 4.1 (Continued)

Botanical name	Habitat	Plant description	Flowering and fruiting	References
Rhodiola sinuata	Rocky crevices, scree slopes; Kumaon; 10,498-14,107 ft.	It is a glabrous perennial herb with long roots. Stem is erect, glabrous or finely mammillate and about 15 cm long. Leaves are alternate, elliptic to ovate, 5-6 × 2-4 cm with attenuated base and pinnately parted margin. Flowers are bisexual and unequally pentamerous; pedicel is finely mammillate or absent; sepals are narrowly triangular; petals are greenish white, elliptic with entire margin and sub-obtuse apex; stamens are 10 in number, and are antespealous and antepetalous ones; carpels are narrowly lanceolate with short styles; seeds are brown, ovoid-oblong and winged at both the ends.	July to September	[34]
Rhodiola stapfii	Grassland Slopes; NE India (Assam); 9514-16,404 ft.	It is a perennial herb with thick rootstock. Stem is simple, erect and 1-3.5 cm long. Leaves are verticillate, ovate with narrow base, entire margin, and sub-obtuse apex. Inflorescence is umbelliform cymes, which is shorter than pedicels. Flowers are unisexual, unequally pentamerous with 1–1.3 cm pedicel; sepals are linear- triangular, 2.5–3.5 mm with connate base and obtuse apex; petals are red, ovate to oblong with 2.5 × 1.7 mm margins; carpels are ovoid-lanceolate with connate base; seeds are brown, oblong, and 0.5-0.8 mm.	August to September	[29]

TABLE 4.1 *(Continued)*

Botanical name	Habitat	Plant description	Flowering and fruiting	References
Rhodiola tibetica	Stony slopes, Sides on ravines; Kashmir and Himachal Pradesh; 13,451-17,716 ft.	It is a glabrous perennial herb with thick and branched rootstock. Stem is simple, smooth, round, reddish and cylindrical. Leaves are alternate, longer than the internodes, sessile, lanceolate to oblong, glaucous, pale on back and about 3 cm long. Inflorescence is terminal and 3-6 cm across with few bracts at the base of branches. Flowers are dark purplish; female flowers have saucer shaped calyx with long-triangular lobes; petals are acute, lanceolate, dark purplish, nearly twice the sepals and dark purplish; carpels are erect with purple tip and short divergent styles.	June to July	[15]
Rhodiola wallichiana	Forests, Rocks on slopes; Kashmir and Sikkim; 8202-12,467 ft.	It is a perennial herb with thick and branched rootstock. Stem is erect and 17-40 cm long. Leaves are numerous, sessile, lanceolate to linear-oblanceolate with attenuated ends and serrated margins. Inflorescence is terminal corymbiform. Flowers are bisexual and unequally pentamerous; sepals are linear with obtuse apex; petals are reddish to yellowish white with obtuse apex; stamens are 10 in number with slightly dilated apex; carpels are erect and ovoid.	July to September	[30]

TABLE 4.1 *(Continued)*

Botanical name	Habitat	Plant description	Flowering and fruiting	References
Rhodiola humilis	Alpine meadow; Sikkim; 12,795-14,763 ft.	It is a glabrous herbaceous perennial with thick and elongated rootstock. Stem is erect, round and 6-13 cm long. Leaves are alternate, sessile, linear-elliptic, 5 × 1 mm with entire margin. Flowers are unequally pentamerous; sepals are ovate-oblong, 3 × 1.5 mm with acute to obtuse apex; petals are ovate, 2.5-3 mm with attenuate apex.	July to September	[15]

TABLE 4.2 Phytochemical and Pharmacological Properties of Selected *Rhodiola* sp.

Botanical name	Plant part	Phytochemical properties	Pharmacological properties	References
Rhodiola Bupleuroides	Roots and rhizomes	Rhobupcyanoside-B 4'-methoxyherbacetin Herbcetin-8-O-α-L-ramnopyranoside Herbcetin-8-methoxy Rhodiosin Quercetin-7-O-β-D-glucopyranoside 3,5,7,8,4'-pentahydroxy-3'-methoxyflavone	α-Glucosidase inhibitory activity	[36]
Rhodiola Crenulata	Roots	Trans-caffeic acid p-Tyrosol Salidroside Creoside Icariside Triandrin Vimalin Rhodiosin	Stress protective, anti-oxidative	[21, 27]
Rhodiola fastigata	Roots	p-Tyrosol Salidroside Daucosterol Cedrusin β-Sitosterol Fastigitin-A	Anti-fatigue, anti-anoxia	[23]
Rhodiola heterodonta	Rhizomes	Salidroside p-Tyrosol	Anti-hypoxic	[13]

TABLE 4.2 *(Continued)*

Botanical name	Plant part	Phytochemical properties	Pharmacological properties	References
Rhodiola imbricata	Roots	Rhodiocyanoside-A Viridoside Mongrhoside Heterodontoside 1-Tricosanol Campesterol b-Fenchyl alcohol d-Tocopherol Thujone Ascaridole Borneol	Anti-microbial, anti-spasmodic, CNS-stimulant, hepato-protective, analgesic	[32]
Rhodiola quadrifida	Under-ground parts	Rosavin Rosarin Rosin Salidroside p-Tyrosol	Adaptogenic, antioxidant	[2]
Rhodiola wallichiana		Salidroside p-Tyrosol Scaphopetalone Berchemol Rhobucyanoside A Chavicol-4-O-β-D apiofuranosyl-(1-6)-O-β-D-glucopyranoside	Anti-tumor metastatic	[31]

In Tibetan medical system, shoots of *Rhodiola heterodonta* are used to treat cold, cough, and lung infection[17], and leaves are used to prepare tantur by local peoples. The bark of *Rhodiola himalensis* is used against tooth related infections.[18]

4.4 PHYTOCHEMISTRY AND PHARMACOLOGICAL PROPERTIES OF SOME *RHODIOLA* SP.

The genus *Rhodiola* is known for having diverse pharmacological properties because of the presence of high content of phytochemicals with potential medicinal applications. Some of the pharmacological properties of some *Rhodiola* species indigenous to India are documented in Table 4.2.

4.5 SUMMARY

Rhodiola is one of the most important plants used in many traditional as well as modern medicines in Asia, Europe, North America, and Alaska. These plants are also found in India (Himalayan Belt), Tibet, China, and Mongolia. It contains high value medicinal compounds (Salidroside, Rosin, and its derivatives) having therapeutic properties for the treatment of high-altitude sickness, depression, cardiovascular, neuronal, liver, and skin disorders as well as epidemic diseases, for example, edema of limbs, injuries, and burns. These compounds also act as immunostimulant, radioprotective, adaptogenic, anti-fatigue, anti-depressant, anti-oxidant, anti-aging, anti-hypoxia, anti-toxic, and anti-cancerous compounds. Thus, the present overview depicts the morphological descriptors and pharmacological properties of *Rhodiola* species, which could serve as a blueprint for screening out different species of *Rhodiola* so that it could be effectively utilized and conserved in a sustainable manner.

KEYWORDS

- **adaptogenic**
- **anti-aging**
- **anti-cancerous**
- **anti-depressant**
- **anti-fatigue**
- **anti-hypoxia**
- **anti-oxidant**
- **Himalaya**
- **immunostimulant**
- **phytomedicine**
- **radioprotective**
- **rosavin**
- **rosin**
- **salidroside**

REFERENCES

1. Alm, T. Ethnobotany of *Rhodiola Rosea* (Crassulaceae) in Norway. *Sida* **2004,** *21* (1), 324–344.
2. Altantsetseg, K.; Przybyl, J. L.; Węglarz, Z.; Geszprych, A. Content of Biologically Active Compounds in Rose Root (*Rhodiola* sp.) Raw Material of Different Derivation. *Herba Polonica* **2007,** *53* (4), 20–26.
3. Biswal, S.; Sharma, D.; Kumar, K.; Nag, T. C.; Barhwal, K.; Hota, S. K.; Kumar, B. Global Hypoxia Induced Impairment in Learning and Spatial Memory is Associated with *Precocious Hippocampal* Aging. *Neurobiol. Learn. Memory* **2016,** *166,* 157–170.
4. Bhardwaj, A. K.; Kapoor, S.; Naryal, A.; Bhardwaj, P.; Warghat, A. R.; Kumar, B.; Chaurasia, O. P. Effect of Various Dormancy Breaking Treatments on Seed Germination, Seedling Growth and Seed Vigor of Medicinal Plants. *Trop. Plant Res.* **2016,** *3* (3), 508–516.
5. Borissova, A. G.; Komarov, V. L. Crassulaceae. In *Flora of the USSR*; Komarov, *V. L.,* Ed.; Publishing House of the USSR Academy of Sciences: Moscow/Leningrad, 1939, vol. 9; pp 8–134, 471–486.
6. Chaurasia, O. P.; Singh, B. *Cold Desert Plants*. Field Research Laboratory, Ministry of Defence, Govt. of India, Leh-Ladakh 901205, Jammu and Kashmir, India; 1996, p 87.
7. Chaurasia, O. P.; Ahmed, Z.; Ballabh, B. *Ethnobotany and Plants of Trans-Himalaya*;: Satish Serial Publishing House: Delhi, 2007; p 132.

8. Chawla, R.; Jaiswal, S.; Kumar, R.; Arora, R.; Sharma, R. K. Himalayan Bioresource *Rhodiola imbricata* as a Promising Radioprotector for Nuclear and Radiological Emergencies. *J. Pharm. Bioall. Sci.* **2010**, *2* (3), 213–219.

9. Choudhary, K.; Singh, M.; Pillai, U. Ethnobotanical survey of Rajasthan: An update. *Am.-Eur. J. Bot.* **2007**, *1* (2), 38–45.

10. Edgeworth, M. P. Descriptions of Some Unpublished Species of Plants from North -Western India. *Trans. Linn. Soc. London* **1846**, *20* (1), 23–47.

11. Gupta, V.; Lahiri, S. S.; Sultana, S.; Tulsawani, R. K.; Kumar, R. Anti-Oxidative Effect of *Rhodiola Imbricata* Root Extract in Rats During Cold, Hypoxia and Restraint (C-H-R) Exposure and Post-Stress Recovery. *Food Chem. Toxicol.* **2010**, *48* (4), 1019–1025.

12. Gupta, S.; Bhoyar, M. S.; Kumar, J.; Warghat, A. R.; Bajpai, P. K.; Rasool, M.; Srivastva, R. B. Genetic Diversity Among Natural Populations of *Rhodiola Imbricate* Edgew from Trans-Himalayan Cold Arid Desert Using Random Amplified Polymorphic DNA (RAPD) and Inter Simple Sequence Repeat (ISSR) Markers. *J. Med. Plants Res.* **2012**, *6* (3), 405–415.

13. Grace, M. H.; Yousef, G. G.; Kurmukov, A. G.; Raskin, I.; Lila, M. A. Phytochemical Characterization of an Adaptogenic Preparation from *Rhodiola Heterodonta. Nat. Prod. Commun.* **2009**, *4* (8), 1053–1058.

14. Hooker, J. D. Rubiaceae: Genera Plantarum Part II. In *The Flora of British India*; Bentham, G., Hooker, J. D., Eds.; Reeve & Co.: London, 1873; p 119.

15. Hooker, J. D.; Thomson. J. *Flora Indica.* London: W. Pamplin; 1855; vol. 1; p 235.

16. Léveillé, H. Gaultheria Laxiflora Diels. *Repertorium Specierum Novarum Regni Vegetabilis* (Vegetable Kingdom Report of the New Species) **1913**, *12*, 283–287.

17. Kala, C. P. Medicinal Plants of the High Altitude Cold Desert in India: Diversity, Distribution and Traditional Uses. *Int. J. Biodiver. Sci. Manag.* **2006**, *2* (1), 43–56.

18. Kumar, M.; Paul, Y.; Anand, V. K. An Ethnobotanical Study of Medicinal Plants Used by the Locals in Kishtwar, Jammu and Kashmir, India. *Ethnobotan. Leaf.* **2009**, *13*, 1240–1256.

19. Kumar, R.; Tayade, A. B.; Chaurasia, O. P.; Sunil, H.; Singh, S. B. Evaluation of Anti-Oxidant Activities and Total Phenol and Flavonoid Content of the Hydro-Alcoholic Extracts of *Rhodiola* sp. *Pharmacog. J.* **2010**, *2* (11), 431–435.

20. Kumar, K.; Sharma, S.; Vashishtha, V.; Bhardwaj, P.; Kumar, A.; Barhwal, K.; Hota, S. K.; Malairaman, U.; Singh, B. *Terminalia Arjuna* Bark Extract Improves Diuresis and Attenuates Acute Hypobaric Hypoxia Induced Cerebral Vascular Leakage. *J. Ethnopharmacol.* **2016**, *180*, 43–53.

21. Nakamura, S.; Li, X.; Matsuda, H.; Yoshikawa, M. Bioactive Constituents from Chinese Natural Medicines: Chemical Structure of Acyclic Alcohol Glycoside from the Roots of *Rhodiola crenulata. Chem. Pharmaceut. Bull.* **2008**, *56* (4), 536–540.

22. Lidén, M.; Bharali, P.; Das, A. K. *Rhodiola Sedoides* (Crassulaceae) - New Species from Arunachal Pradesh, India. *Annales Botanici Fennici* **2016**, *53*, 106–108.

23. Liu, H. J.; Xu, Y.; Liu, Y. J.; Liu, C. Z. Plant Regeneration From Leaf Explants of *Rhodiola Fastigiata. In Vitro Cell. Dev. Biol.-Plant* **2006**, *42* (4), 345–347.

24. Pallas, R.; Hooker, J. D.; Thompson, J. Rosen Root. *Flora Br. India* **1876**, *2*, 412–418.

25. Panossian, A.; Wikman, G. Rosenroot (*Rhodiola Rosea*): Traditional Use, Chemical Composition, Pharmacology and Clinical Efficacy. *Phytomedicine* **2010**, *17* (7), 481–493.

26. Prager, L. R. An Account of the Genus Sedum as Found in Cultivation; Wheldon & Wesley: Frankfurt, Germany, 1967; p 314.

27. Qu, Z. Q.; Zhou, Y.; Zeng, Y. S.; Lin, Y. K.; Li, Y.; Zhong, Z. Q.; Chan, W. Y. Protective Effects of a *Rhodiola Crenulata* Extract and Salidroside on Hippocampal Neurogenesis Against Streptozotocin-Induced Neural Injury in the Rat. *PLoS One* **2012,** *7* (1), 29641

28. Raymond, H.; Ohba, H. *Rhodiola Prainii. J. Jap. Bot.* **1976,** *51*, 386.

29. Raymond, H. *Sedum Prainii. Bull. Soc. Bot. France* **1909,** *56*, 566.

30. Si, H. F. *Rhodiola Taohoensis. Acta Phytotaxonom. Sin. Additament.* **1965,** *1*, 121.

31. Song, Y. Ferulic Acid Ester from *Rhodiola Wallichiana* var. Cholaensis (Crassulaceae). *Nat. Prod. Res.* **2017,** *2017*, 1–8.

32. Tayade, A. B.; Dhar, P.; Kumar, J.; Sharma, M.; Chuahan, R. S.; Chaurasia, O. P.; Srivastava, R. B. Chemometric Profile of Root Extracts of *Rhodiola Imbricate* Edgew with Hyphenated Gas Chromatography Mass Spectrometric (HGC-MS) Technique. *PLoS One* **2013,** *8* (1), 1–15.

33. Yuan, C. *Sedum Coccineum* Royle, Ill. *Bot. Himal. Mts.* **1835,** *1*, 223.

34. Yi, C. *Sedum Heterodontum. J. Proceed. Linn. Soc. Bot.* **1858,** *2*, 95.

35. Wallich, N. A Numerical List of Dried Plants; East India Company's Museum (Wallich's Catalogue): London, 1847; pp. 8234–8521.

36. Wang, H.; Dong, L.; Ge, J. Q.; Lan, X. Z.; Lio, Z. H.; Chen, M. Rhobup Cyanoside – B from *Rhodiola Bupleuroides. J. Asian Nat. Prod. Res.* **2016,** *18* (11), 1108–1114.

CHAPTER 5

POTENTIAL OF *PIPER* GERMPLASM AGAINST PATHOGENIC BACTERIA: TROPICAL BAY ISLANDS IN INDIA

CHINTHAMANI JAYAVEL, AJIT ARUN WAMAN,
SARAVANAN KANDASAMY, and POOJA BOHRA

ABSTRACT

This study deals with morphological characterization, determination of photosynthetic pigments, and testing of antibacterial activities of 10 *Piper* genotypes from Andaman and Nicobar Islands. Results revealed significant differences among the genotypes under study. Plants of *P. colubrinum*-2 had longer, heavier leaves and higher petiole thickness, while that of *P. sarmentosum*-3 had broadest leaves and widest petioles. Dry matter content of the genotypes varied from 18 to 40%. Total chlorophyll content in the leaves varied from 0.60 to 6.72 mg/g. Wide variations were also observed in antibacterial activity of ethanolic extracts of genotypes at two doses (80 and 100 µL) against seven bacterial pathogens. *P. colubrinum*-1 and *P. colubrinum*-2 were effective in inhibiting the growth of *Staphylococcus aureus* and results were comparable with positive control. In case of *Pseudomonas fluorescens*, extracts of all genotypes were effective as their zone of inhibition was found to be statistically on par with the control chloramphenicol. Although moderate level of inhibition was observed for rest of the pathogens, yet it remained much lower than the inhibition observed for positive control. Authors affirm that while working on a new species, chance screening of an ineffective genotype may completely eliminate the candidature of that species for further experiments. Hence, while screening any new species for its antimicrobial potential, using the available intraspecific diversity of that species can help in ascertaining the actual potential of the species and identification of superior types as observed in this research.

5.1 INTRODUCTION

The tropical Andaman and Nicobar Islands are a group of 572 islands and isles in the Bay of Bengal and are known to harbor variety of flora and fauna. However, the islands are vulnerable to natural and manmade disasters, which have adversely impacted both native and introduced flora to a great extent.[40] Systematic efforts are required to characterize the diversity of important taxa, which may be beneficial for the mankind in solving various challenges. *Piper* is an economically important genus of the botanical family Piperaceae with more than 2000 species distributed in different parts of the world.[1]

Most of the *Piper* species are known for their culinary and medicinal uses. About eight *Piper* species have been reported from the islands and they are considered as the centers of origin for *P. betle* L. because of distinct morphotypes reported from these islands.[41] Considerable interspecific diversity is noticed for the genus in the islands and their potential should be studied in detail.[40] Further, evaluation of the intraspecific diversity in a species may help in identification of more potent types with desired traits.[10,48]

Pathogens have been evolving ever since their origin. However, recent era has witnessed rapid development of new strains causing variety of diseases. At present, these pathogens are mainly being controlled by using antibiotics, which have caused adverse effects in the form of host toxicity and antibiotic resistance.[22] Herbal drugs are considered to be safe as they are known to overcome the toxicity caused by pathogens without any side effects.[39] Also, the problem of drug resistance could be overcome by using alternative drugs.[11] Hence, screening of potential plant sources for obtaining novel herbal drugs could be of great practical utility.[50]

Several plant species are known to possess antimicrobial properties and are considered as a boon in the wake of increasing incidences of multidrug resistant microbes.[6] These species can be used in the preparation of drug formulations against pathogens of bacterial, fungal, and protozoal origin.[7] Spices and medicinal plants are among the most common natural sources that possess antimicrobial activity against variety of pathogens.[49] Studies have claimed that antibacterial activity of spices is attributed to the presence of active ingredients including phenolic compounds in them.[15] Though antibacterial activity against both Gram-positive and Gram-negative species has been documented in some of the *Piper* species,[52] the unexplored germplasm of different species need to be screened for obtaining superior genotypes.

This chapter focuses on the characterization of 10 *Piper* germplasm collected from different parts of the Bay islands for morpho-biochemical characteristics and their evaluation against seven pathogenic bacteria.

5.2 MATERIAL AND METHODS

5.2.1 MORPHOLOGICAL CHARACTERIZATION

Piper species (nine genotypes) were collected from the South Andaman Island, Middle Andaman Island, and North Andaman Island, while *P. chaba* Hunt. was obtained from ICAR-Indian Institute of Spices Research, Kozhikode, India. The species were maintained in the germplasm block of the Division of Horticulture and Forestry at ICAR—Central Island Agricultural Research Institute, Port Blair. Leaves were collected from 10 genotypes, viz., *Piper chaba, P. colubrinum* -1 and 2, *P. betle, P. sarmentosum* -1 to 5, and *P. nigrum* L. Table 5.1 indicates common names of these species in different languages, major active ingredients, and antibacterial activities reported earlier. The observations on morphological parameters were replicated 10 times and these were: leaf length (cm), leaf width (cm), petiole length (cm), petiole width (mm), petiole thickness (mm), leaf weight (g), and petiole weight (mg). For determining dry matter content (%), 1 g of leaf tissue was weighed accurately and samples were kept in hot air oven maintained at 60°C till constant weight was obtained. The dry matter content was calculated using the following formula:

$$\text{Dry matter content}\,(\%) = \frac{\text{Dry weight}}{\text{Fresh weight}} \times 100 \tag{5.1}$$

5.2.2 DETERMINATION OF CHLOROPHYLL CONTENT

Four physiologically active leaves of each genotype were collected from the germplasm block and were washed thoroughly in running tap water. Leaves were cut into small pieces, mixed well, and 100 mg of each sample was used for grinding using mortar and pestle. After thorough grinding, 1 mL of chilled acetone (80%) was added to the sample and supernatant was collected in centrifuge tubes. The procedure was repeated till the tissues were devoid of green color and the final volume was made up to 5 mL using the same solvent. Samples were centrifuged at 5000 rpm for 5 min at 4°C and 50 µL of the supernatant was transferred into a cuvette using micropipette. Absorbance was recorded at 645 nm and 663 nm using Biospectrometer (Eppendorf, Germany). The amount of Chl-a, Chl-b, and total chlorophyll in the samples were calculated using the following formulae[37]

TABLE 5.1 *Piper* Species Under Investigation in This Chapter.

Species and their vernacular names	Active ingredients	Refs.	Selected antibacterial activity	Refs.
P. betle: Betel leaf (E), Paan (H), Tanbul (A), Ju Jiang (C), Sirih (I, M), Pelu (T), Traukhong (V)	Flavonoids, amides, pyrones	[46]	*Staphylococcus aureus, Escherichia coli, Pseudomonas aeruginosa, Bacillus cereus; Klebsiella pneumoniae*	[19, 47]
P. chaba: Javanese long pepper, Balanese pepper (E), Chavi, Chavya (H), Jia bi bo (C), Ladapanjang (M), Dee Plee (T), La lot (V)	Amides, chalcones, flavonoids, lignans, terpenoids, pyrones, steroids	[31, 38]	*Bacillus cereus; Klebsiella pneumoniae*	[47]
P. colubrinum: Brazilian wild pepper (E), Brazilian Hippali (H)	Flavone	[4]	–	–
P. nigrum: Pepper (E), Golmirch (H), FilFil, FulFul (A), Heihujiao (C), Maricahitam (I), Ladapadi (M), Phrikthai (T), Trieu (V)	Piperine, piperolein B	[23]	*Bacillus subtilis, Escherichia coli, Pseudomonas fluorescens*	[30, 45]
P. sarmentosum: Wild betel leaf (E), Jia Ju (C), Sirih Tanah (I), Chabei, Pokok Kadok (M), Cha phlu (T), La lop (V)	Alkaloids, amides, flavonoids, lignans, phenylpropanoids	[12, 18]	*Staphylococcus aureus, Pseudomonas aeruginosa*	[52]

Legend:
E, English; H, Hindi; C, Chinese; A, Arabic; I, Indonesian; M, Malayan; T, Thai; V, Vietnamese.

$$\text{Chl a}\left(\text{mg / g tissue}\right) = 12.7\left(A663\right) - 2.69(A645)\times\frac{V}{W\times1000} \qquad (5.2)$$

$$\text{Chl b}\left(\text{mg / g tissue}\right) = 22.9\left(A645\right) - 4.68(A663)\times\frac{V}{W\times1000} \qquad (5.3)$$

$$\text{Total Chl}\left(\text{mg / g tissue}\right) = 20.2\left(A645\right) + 8.02(A663)\times\frac{V}{W\times1000} \qquad (5.4)$$

where, V is the total volume made up; and W is weight of the tissue taken.

5.2.3 PREPARATION OF EXTRACTS

Leaf samples of 10 germplasm were collected, shade dried, and powdered using an electric blender. Samples were extracted using cold percolation technique, for which 1 g of powder was transferred to glass vial and 10 mL of 80% ethanol (v/v) was added. Vials were then maintained in dark for 48 h with intermittent shaking. The extracts were then filtered using syringe filter (0.2 µ) and the solvent was evaporated to dryness at 40°C. The samples were re-dissolved in 10 mL of 80% ethanol and used for determining antibacterial activity.

5.2.4 MICROORGANISMS AND PREPARATION OF INOCULUM

Pure cultures of seven bacteria (Table 5.2), viz., *Aeromonas hydrophilla* (ATCC35654), *Staphylococcus aureus* (ATCC 25923), *Escherichia coli* (ATCC 4157), *Streptococcus pneumonia* (ATCC 49619), *Vibrio algino-lyticus* (ATCC 17749), *Edward siellatarda* (ATCC 15947), and *Pseudomonas fluorescens* (ATCC 13525) were procured from Microbiologics, USA, and used in the present investigation. One loopful of culture was subcultured into 10 mL nutrient broth and incubated at 37°C in the shaking incubator (100 rpm) for 24 h.

5.2.5 PREPARATION OF PLATES FOR ANTIBACTERIAL ACTIVITY

Antibacterial activity of the ethanolic extracts of *Piper* samples was determined using agar well diffusion method.[16] For this, Mueller Hinton agar (HiMedia, Mumbai, India) plates were prepared and 50 µL of test culture was transferred aseptically to each plate and spread using a sterile L-rod followed by 30 min incubation. Using sterile cork borer (5.5 mm in diameter), four wells were made in each plate and 80 or 100 µL of prepared extract, as per the treatment, was inoculated in each well. Streptomycin (1 mg/mL) and chloramphenicol (1 mg/mL) were used as positive controls, whereas 80% ethanol was used as the negative control. Plates were then incubated at 37°C in an orbital shaking incubator (Remi, India). Each sample was maintained in triplicate to minimize the experimental error. Zone of inhibition in response to the challenging was measured after 24 h.

TABLE 5.2   Seven Bacterial Strains Used in Present Investigation and Their Characteristics.

Bacterial strain	Characteristics	Pathogenesis	Ref.
Aeromonas hydrophilla	Gram negative, rod shaped	Causes haemorrhagic septicaemia, and red pest disease in fish	[29, 36]
Edwardsiella tarda	Gram negative, rod shaped	Causes Edwardsiellosis, Edwardsiella septicaemia, and emphysematous putrefactive disease in fish	[29, 36]
Escherichia Coli	Gram negative, rod shaped	Causes urinary tract infection, neonatal meningitis, enteric infections, bacteremia, nosocomial pneumonia, peritonitis, cellulitis, and infectious arthritis in humans	[3]
Pseudomonas fluorescens	Gram negative, rod shaped	Causes abdominal dropsy, septicaemia, and fin rot disease in fish	[29, 36]
Staphylococcus aureus	Gram positive, coccus shaped	Causes endocarditis, osteomyelitis, septic arthritis, pleuropulmonary infections, meningitis, toxic shock syndrome, scalded skin syndrome	[43]
Streptococcus pneumoniae	Gram positive, lancet shaped	Causes pneumonia, otitis media, bacteremia, and meningitis in humans	[25]
Vibrio alginolyticus	Gram negative, rod shaped	Causes vibriosis in fish	[29]

5.3   RESULTS AND DISCUSSION

Germplasm is the basic requirement for taking up any crop improvement program as it provides an opportunity for identification of useful genotypes for direct and indirect utilization. Andaman and Nicobar group of islands are known to harbor considerable diversity of flora of economic importance and *Piper* is one such genus, which is found naturally in different parts of the islands.[40] Exploration in the islands and collection of species from other sources has resulted in the establishment of germplasm block of *Piper* species at ICAR-CIARI, Port Blair. Ten interspecific and intraspecific germplasm of *Piper* were selected for the present investigation.

5.3.1 MORPHOLOGICAL CHARACTERIZATION

Data in Table 5.3 revealed considerable variations for all the parameters in the studied samples. In general, except for leaf length, wherein *P. betle* was found to have shortest leaves, values for all other parameters were lowest in *P. chaba*. The *P. colubrinum*-2 had highest values of leaf length, petiole thickness, and leaf weight. Leaf width and petiole width were highest in *P. sarmentosum*-3, while petiole length was highest in *P. betle*. Variability was noticed for leaf length (8.97–18.17 cm), leaf width (4.06–10.7 cm), petiole length (0.76–6.22 cm), petiole width (1.53–3.92 mm), petiole thickness (1.33–3.19 mm), and fresh leaf weight (0.67–3.29 g).

Characterization is the description of important traits of a genotype that is highly heritable and is expected to express in all the environments. Morphological parameters are the important component in the characterization process that is used to establish taxonomic identity of a genotype and for conservation of rare species.[51] Morphological characterization is considered as the first step[5] before taking up detailed molecular and biochemical studies. The commonly adopted, simple, and reliable morphological traits used to differentiate among the genotypes are the leaf parameters.[24] Recent report[10] on antidiabetic plant, *Gymnema sylvestre*, suggested that leaf parameters could be used for characterization of genotypes involving intraspecific variants collected from different geographical regions. Hence, the differences in these parameters observed during present investigation could help to establish identity of the studied samples.

Leaf-related parameters can also help in studying the physiology of a genotype. For example, leaf size is directly proportional to its primordia from which it emerges out.[28] Larger leaf area in a genotype may possibly be associated with higher photosynthetic activities than a genotype of same species with smaller leaf area. Further, the gaseous exchange and water movement in the system has close association with the morphology of the species. However, studies on a number of other factors could be necessary to support this. Petioles play a key role in shading of leaves, in which longer petioles could inhibit self-shading effects.[42] Hence, during the present investigation, leaf and petiole-related parameters were used for morphological characterization of the *Piper* genotypes. Variations for morphological

TABLE 5.3 Morphological Characterization of the *Piper* Genotypes.

Sample	Leaf length	Leaf width	Petiole length	Petiole width	Petiole thickness	Leaf Weight	Dry matter
	Cm			mm		G	%
P. betle	8.97 ± 0.167	8.31 ± 0.156	6.22 ± 0.192	3.19 ± 0.132	1.92 ± 0.045	1.84 ± 0.074	23.3
P. chaba	11.33 ± 0.261	4.06 ± 0.301	0.76 ± 0.128	1.53 ± 0.164	1.33 ± 0.084	0.67 ± 0.048	40.7
*P. colubrinum-*1	17.97 ± 0.460	8.47 ± 0.222	1.18 ± 0.160	2.35 ± 0.107	2.70 ± 0.146	1.98 ± 0.098	32.9
*P. colubrinum-*2	18.17 ± 0.360	9.53 ± 0.256	1.59 ± 0.247	3.17 ± 0.085	3.19 ± 0.082	3.29 ± 0.096	18.1
P. nigrum	9.97 ± 0.544	8.66 ± 0.241	4.20 ± 0.082	2.23 ± 0.544	2.08 ± 0.241	1.56 ± 0.082	21.2
P. sarmentosum- 2	9.90 ± 0.370	8.93 ± 0.303	4.01 ± 0.226	3.50 ± 0.251	1.61 ± 0.071	1.05 ± 0.061	23.7
P. sarmentosum- 3	12.97 ± 0.690	10.70 ± 0.631	4.12 ± 0.403	3.92 ± 0.396	1.84 ± 0.101	1.57 ± 0.149	24.8
P. sarmentosum- 4	9.83 ± 0.491	8.55 ± 0.409	3.40 ± 0.258	2.99 ± 0.278	1.41 ± 0.073	0.94 ± 0.119	20.3
P. sarmentosum- 5	10.52 ± 0.295	9.17 ± 0.133	3.80 ± 0.271	3.63 ± 0.199	1.58 ± 0.056	1.07 ± 0.034	20.5
P. sarmentosum -1	12.13 ± 0.444	10.65 ± 0.545	4.52 ± 0.454	3.69 ± 0.335	1.80 ± 0.078	1.67 ± 0.147	36.2

Note: The values in this table are mean ± standard error of 10 replicates.

parameters among different genotypes are in line with the earlier report[5] on the characterization of the diversity of *Piper* spp. from North Eastern India.

In case of dry matter content, highest recovery (40.7%) was observed in *P. chaba*, while it was the lowest (18.1%) in *P. colubrinum*- 2. Among the genotypes of *P. sarmentosum,* the biomass content varied from 20.3 to 36.2%, while it varied from 18.1 to 32.9% in *P. colubrinum* collections. About 23.3% and 21.2% biomass was observed in *P. betle* and *P. nigrum*, respectively. The reports on leaf dry matter content in *Piper* species are scarce, except for a report by Parthasarathy et al.[32] on *P. nigrum* genotypes. They suggested that the dry matter content varied between 19 and 21% among the accessions collected from different parts of Karnataka and Kerala, and interestingly almost 90% of the samples had similar range. Considering the large variations among the intraspecific collections of *P. sarmentosum* and *P. colubrinum* observed during the present investigation, detailed research studies could possibly help in establishing their correlation with traits of economic importance.

5.3.2 DETERMINATION OF CHLOROPHYLL CONTENT

Chlorophyll is a vital component for photosynthesis, which is comprised of Chl-a (primary pigment) and Chl-b (accessory pigment). Significant differences were noticed among the samples studied for total chlorophyll content (Table 5.4). In general, total chlorophyll content varied from 0.60 to 6.72 mg/g. Wide variations were noticed among genotypes of *P. sarmentosum* as the values varied between 0.90 and 6.72 mg/g in this species. In case of Chl-a and Chl-b, *P. sarmentosum*-4 was found to contain the highest amount of pigments of 3.31 mg/g and 3.42 mg/g, respectively. *P. colubrinum*-2 was found to contain lowest amount of Chl-a (0.24 mg/g) and Chl-b (0.37 mg/g) among all the genotypes studied. In general, concentration of Chl-b was comparatively higher than that of Chl-a, except in case of *P. sarmentosum*-1.

Concentration of chlorophyll pigment in a genotype is influenced by number of factors including stress level,[33] nutritional content of the tissue,[26] pollution and environmental stress,[44] and physiological status of the plant.[9] Variations in chlorophyll content in *Piper* species are also reported to be due to soil and micro-environmental factors, stage of plant growth, and sex of the genotype.[20,32] However, during present investigation, all the genotypes were grown with uniform management practices and hence, the differences in the pigment content could be attributed to the differences in their genetic make-up. Degradation of chlorophyll[17] and variation of chlorophyll content[14] is mainly attributed to the presence of primary enzyme, viz., Chlorophyllase. The differences observed during present investigation could possibly be due to varied levels of Chlorophyllase enzyme in the tissues of different genotypes. Findings of the present study are supported by an earlier study of 22 genotypes of *P. nigrum*.[32]

TABLE 5.4   Estimation of Chlorophyll (Chl) Content in Leaf samples of *Piper* Genotypes.

Sample	Concentration (mg/g tissue)		
	Chl-a	Chl-b	Total Chl
P. betle	0.34 ± 0.050	0.57 ± 0.075	0.90 ± 0.120
P. chaba	0.80 ± 0.260	1.39 ± 0.210	2.18 ± 0.470
P. colubrinum-1	1.12 ± 0.020	1.47 ± 0.155	2.59 ± 0.135
P. colubrinum-2	0.24 ± 0.211	0.37 ± 0.312	0.60 ± 0.523
P. nigrum	1.47 ±0.080	1.56 ± 0.050	3.03 ± 0.130
P. sarmentosum -1	0.83 ± 0.005	0.82 ± 0.037	1.58 ± 0.025
P. sarmentosum- 2	0.55 ± 0.062	0.73 ± 0.073	1.17 ± 0.113
P. sarmentosum- 3	1.45 ± 0.235	1.49 ± 0.335	2.94 ± 0.575
P. sarmentosum- 4	3.31 ± 0.005	3.42 ± 0.360	6.72 ± 0.350
P. sarmentosum- 5	0.50 ± 0.140	0.60 ± 0.140	0.90 ± 0.175

5.3.3   SCREENING FOR ANTIBACTERIAL ACTIVITIES

Spices are the most common natural sources possessing antibacterial activities against many species of bacteria. These activities are mainly attributed to the presence of secondary metabolites including phenolic compounds.[15] Most of these compounds can be effectively extracted using various solvents and used for testing antibacterial activities. Antibacterial activity against both Gram-positive and Gram-negative species has been well documented in a number of *Piper* species.[52] The present study was undertaken to screen the antibacterial activities of ethanolic extracts of 10 *Piper* genotypes at two doses (80 and 100 μL) against seven pathogenic bacteria using agar well diffusion method. In general, all samples were tested for its activity up to 24 h (Tables 5.5 and 5.6).

In case of *Aeromonas hydrophila*, results revealed that in all the extracts, 100 μL dose showed higher zone of inhibition than that with 80 μL. Extracts of *P. colubrinum*-2 and *P. sarmentosum*-2 remained ineffective as no inhibition zone was noticed in these treatments at both concentrations. Though some inhibition activity was noticed in the remaining extracts, none of these was statistically comparable with the positive controls (Table 5.5). Earlier reports on this bacterium suggested that *P. betle* had inhibitory effects against *A. hydrophilla* (ATCC 7965) with 11.0 mm inhibition.[2] During present investigation, inhibition zone of 9.58 mm was noticed in *P. betle* genotype, yet it remained much lower than the inhibition observed with the positive controls. Another report also suggested moderate response against the pathogens.[27]

Studies with *Edwardsiella tarda* showed that *Piper* genotypes could inhibit the growth of pathogen to some extent, whereas it remained lower than the inhibition observed for streptomycin (Table 5.5). In *P. colubrinum*-2, no inhibitory activity was seen, while *P. colubrinum*-1 showed inhibition zone at 100 µL. Contradictory reports are available for *P. betle* against the pathogen as some reports suggested its antibacterial activity,[2,21] while poor response was observed in other,[27] These variations could be attributed to the variation in the genotypes used and the strains/virulence of the pathogens employed.

Results regarding *Escherichia coli* revealed dose-dependent response (Table 5.5). However, none of the extract could effectively inhibit the growth of the bacteria as their inhibition zones remained significantly lower than those observed with the positive controls. Extracts of *P. chaba* and two genotypes of *P. sarmentosum*-3 and 4 failed to show any inhibitory activity against the pathogen. Earlier reports on *E. coli* suggested that ethanolic extract of *P. sarmentosum*[52] showed inhibition activity. Noncompliance during the present study could be attributed to differences in the genotypes used during these studies.

In case of *Pseudomonas fluorescens*, results revealed that extracts of all the genotypes were effective as their zone of inhibition was statistically on par with the control chloramphenicol. Low antibacterial activity was recorded in extracts of *P. chaba* and *P. betle*. However, zone of inhibition of all the extracts remained much lower than the streptomycin control (Table 5.5). Low inhibitory activities of *P. betle* against *P. fluorescence* have already been recorded.[27]

In *Vibrio alginolyticus*, screening of extracts revealed that none of the extracts was effective enough to inhibit its growth, when compared with the positive controls. Extracts of *P. chaba* and *P. sarmentosum*-1, 4, and 5 could inhibit the pathogen growth only at higher concentration of 100 µL (Table 5.5). Ethanolic extract of black pepper showed 14 mm inhibition zone against *V. alginolyticus*,[34] while it was 9.58 mm in present study. Extracts of *P. betle* showed 12.5 mm zone of inhibition[2] against *V. alginolyticus* (ATCC 17749) as compared to 7.75 mm as observed in the present study.

TABLE 5.5 Antibacterial Activity of 10 *Piper* Leaf Extracts Using Two Inoculum Volumes Against Five Gram-Negative Bacteria After 24 h of Incubation.

Sample	Aeromonas hydrophilla		Edwardsiella tarda		Escherichia coli		Pseudomonas Fluorescens		Vibrio alginolyticus	
	80 µL	100 µL	80 µL	100 µL	80 µL	100 µL	80 µL	100 µL	80 µL	100 µL
P. Betle	7.38 ± 0.125	9.58 ± 0.960	7.92 ± 0.417	8.25 ± 0.382	—	6.25 ± 0.250	14.67 ± 0.726	14.67 ± 1.202	7.08 ± 0.083	7.75 ± 0.382
P. chava	−0	8.83 ± 1.013	—	6.33 ± 0.333	—	—	10.83 ± 0.600	12.83 ± 1.167	—	6.83 ± 0.167
P. colubrinum-1	7.42 ± 0.083	8.50 ± 0.289	—	6.25 ± 0.204	8.17 ± 0.601	10.17 ± 0.441	16.00 ± 0.000	18.50 ± 0.764	7.00 ± 0.000	7.00 ± 0.000
P. colubrinum-2	—	—	—	—	9.83 ± 0.441	11.00 ± 0.764	16.17 ± 0.167	16.83 ± 1.364	7.67 ± 0.333	8.00 ± 0.289
P. nigrum	—	7.50 ± 1.155	6.42 ± 0.300	7.58 ± 0.300	—	6.25 ± 0.250	14.50 ± 1.607	15.50 ± 1.443	8.50 ± 0.764	9.58 ± 0.583
P. sarmentosum-1	—	8.00 ± 0.577	9.42 ± 1.402	11.17 ± 2.210	—	6.00 ± 0.000	16.33 ± 2.667	17.83 ± 0.441	—	7.17 ± 0.441
P. sarmentosum-2	—	—	7.25 ± 0.250	7.92 ± 0.083	—	6.33 ± 0.333	20.50 ± 4.163	19.67 ± 2.619	6.00 ± 0.000	7.58 ± 0.464
P. sarmentosum-3	—	7.50 ± 0.000	7.00 ± 0.000	8.83 ± 0.441	—	—	15.17 ± 1.856	16.67 ± 1.922	6.50 ± 0.408	7.00 ± 0.000
P. sarmentosum-4	—	7.50 ± 0.500	7.00 ± 0.577	7.58 ± 0.300	—	—	13.83 ± 0.601	17.00 ± 1.323	—	6.88 ± 0.510
P. sarmentosum-5	—	7.75 ± 0.750	6.00 ± 0.000	7.33 ± 0.333	6.00 ± 0.000	6.00 ± 0.000	15.50 ± 1.607	17.00 ± 2.566	—	8.25 ± 0.612
Streptomycin	16.17 ± 0.441	17.08 ± 0.583	20.50 ± 1.041	21.50 ± 2.082	21.17 ± 0.601	21.33 ± 0.441	30.00 ± 0.000	31.50 ± 0.866	20.67 ± 0.333	22.17 ± 0.167
Chloramphenicol	24.83 ± 0.167	26.33 ± 0.167	—	—	21.75 ± 0.612	23.88 ± 0.102	19.00 ± 0.408	17.50 ± 0.816	19.83 ± 0.167	21.00 ± 0.764

Note: The values in this table are mean ± standard error of three replicates.

In case of *Staphylococcus aureus*, a major human pathogen, extracts of only four samples were effective as they showed inhibitory activity (Table 5.6) and in all those cases, higher doses of inoculum were more effective. Interestingly, only two collections of *P. colubrinum* could show inhibitory activities (15.67 mm and 15.17 mm), which remained on par with the positive control chloramphenicol. Earlier report suggested antibacterial activity of *P. betle* against *S. aureus*;[8] however, during present investigation, no inhibitory activity was noticed.

Results on antibacterial activities of *Piper* genotypes against *Streptococcus pneumoniae* suggested that though some activity was noticed in seven of the genotypes, none of them was comparable with the zone of inhibition recorded with controls (Table 5.6). Absence of activity was observed in *P. chaba* and *P. sarmentosum*-1 and 5. In other studies, chloroform extracts of *P. betle* have been found to show inhibitory activities (11 mm) against *Streptococcus pneumoniae* by disc diffusion method.[35] Earlier reports suggested that crude extract of *P. sarmentosum* had no activity against *S. pneumoniae*;[13] however, inhibitory activities exhibited by three collections of *P. sarmentosum* during present study indicated that the pathogens are influenced in genotype dependent manner.

TABLE 5.6 Antibacterial Activity of 10 *Piper* Leaf Extracts Using Two Inoculum Volumes Against Two Gram-Positive Bacteria after 24 h of Incubation.

Genotype code	Staphylococcus aureus		Streptococcus pneumonia	
	80 µL	100 µL	80 µL	100 µL
P. sarmentosum-2	–	–	–	7.00 ± 0.500
P. betle	–	–	8.33 ± 0.583	9.83 ± 0.882
P. chaba	–	6.50 ± 0.289	–	–
P. colubrinum-1	15.08 ± 0.220	15.17 ± 0.601	15.33 ± 1.481	16.67 ± 0.726
P. colubrinum-2	14.67 ± 0.441	15.67 ± 0.333	14.33 ± 1.302	16.17 ± 0.601
P. nigrum	–	–	7.25 ± 0.250	8.42 ± 0.300
P. sarmentosum -1	–	–	–	–
P. sarmentosum-3	–	8.58 ± 0.083	7.75 ± 1.750	6.25 ± 0.250
P. sarmentosum-4	–	–	7.17 ± 0.167	8.25 ± 0.000
P. sarmentosum-5	–	–	–	–
Streptomycin	16.33 ± 0.726	17.50 ± 1.041	34.00 ± 1.000	36.17 ± 0.441
Chloramphenicol	13.33 ± 1.667	14.33 ± 0.882	25.67 ± 1.453	27.50 ± 0.764

Note: The values in this table are mean ± standard error of three replicates.

5.4 SUMMARY

The present study deals with morphological and biochemical characterization, and testing the antibacterial activities of 10 *Piper* genotypes from Andaman and Nicobar Islands in India. Results revealed significant differences among the genotypes studied. Plants of *P. colubrinum*-2 had longer, heavier leaves, and higher petiole thickness, while that of *P. sarmentosum*-3 had broadest leaves and widest petioles. Dry matter content of the genotypes varied from 18 to 40%. Amount of total chlorophyll content in the leaves varied from 0.60 to 6.72 mg/g. Wide variations were also observed in antibacterial activity of ethanolic extracts of the studied genotypes at two doses (80 and 100 µL) against seven bacterial pathogens. *P. colubrinum*-1 and 2 were effective in inhibiting the growth of *Staphylococcus aureus* and results were comparable with positive control. In case of *Pseudomonas fluorescens,* extracts of all the genotypes were effective as their zone of inhibition was statistically on par with the control chloramphenicol. Though moderate level of inhibition was observed for rest of the pathogens, it remained much lower than the inhibition observed for positive control. Authors indicate that while working on a new species, chance screening of an ineffective genotype could completely eliminate the candidature of that species for further experiments. Hence, while screening any new species for its antimicrobial potential, using the available intraspecific diversity of that species could help in ascertaining the actual potential of the species and identification of superior types as observed in the present study.

ACKNOWLEDGMENTS

Authors are thankful to the director of ICAR-Central Island Agricultural Research Institute for providing necessary facilities for conducting the experiment. Part of the work was funded under Consortia Research Project on Agrobiodiversity, which is gratefully acknowledged by AAW.

KEYWORDS

- **Andaman Islands**
- **antimicrobial activity**

- **chlorophyll**
- **diversity**
- **ecotypes**
- **genetic resources**
- **Piperaceae**

REFERENCES

1. Airy-Shaw, H. K. *A Dictionary of the Flowering Plants and Ferns*. Cambridge University Press: New York, 1973; p 365.
2. Albert, V.; Ransangan, J. Antibacterial Potential of Plant Crude Extracts against Gram Negative Fish Bacterial Pathogens. *Int. J. Pharmaceut. Res. Biosci.* **2013**, *3*, 21–27.
3. Bachir, R. G.; Abouni, B. *Escherichia Coli* and *Staphylococcus Aureus* Most Common Source of Infection. In *The Battle Against Microbial Pathogens: Basic Science, Technological Advances and Educational Programs*; Méndez-Vilas, A., Ed.; Formatex Research Center: Spain, 2015; pp 637–648.
4. Banerji, A.; Rej, R. N.; Ghosh, P. C. Isolation of N-Isobutyl Deca-Trans-2-Trans-4-Dienamide from *Piper Sylvaticum* Roxb. *Cell. Mol. Life Sci.* **1974**, *30*, 223–224.
5. Chongtham, C.; Thongam, B; Handique, P. J. Morphology Diversity and Characterization of Some of the Wild *Piper* Species of North East India. *Gen. Res. Crop Evol.* **2015**, *62*, 303–313.
6. Cosgrove, S.; Carmeli, Y. The Impact of Antimicrobial Resistance on Health and Economic Outcomes. *Clin. Infect. Dis.* **2003**, *36*, 1433–1437.
7. Cowan, M. M. Plant Products as Anti-Microbial Agents. *Clin. Microbiol. Rev.* **1999**, *12*, 564–582.
8. Datta, A.; Ghoshdastidar, S.; Singh, M. Antimicrobial Property of *Piper Betle* Leaf against Clinical Isolates of Bacteria. *Int. J. Phar. Sci. Res.* **2011**, *2*, 104–109.
9. Demmig-Adams, B.; Adams, W. W. Carotenoid Composition in Sun and Shade Leaves of Plants with Different Life Forms. *Plant Cell Environ.* **1992**, *15*, 411–419.
10. Dhanani, T.; Singh, R.; Waman, A.; Patel, P.; Manivel, P.; Kumar, S. Assessment of Diversity amongst Natural Populations of *Gymnema Sylvestre* from India and Development of a Validated HPLC Protocol for Identification and Quantification of Gymnemagenin. *Indust. Crops Prod.* **2015**, *77*, 901–909.
11. Dimayuga, R. E.; Garcia, S. K. Anti-Microbial Screening of Medicinal Plants From Baja California Sur. Mexico. *J. Ethnopharmacol.* **1991**, *31*, 181–192.
12. Ee, G. C. L.; Lim, C. M.; Rahmani, M.; Shaari, K.; Bong, C. F. G. Alkaloids From *Piper Sarmentosum* and *Piper Nigrum*. *Nat. Prod. Res.* **2009**, *23*, 1416–1423.
13. Fernandez, L.; Daruliza, K.; Sudhakaran, S.; Jegathambigai, R. Antimicrobial Activity of the Crude Extract of *Piper Sarmentosum* against Methicilin-Resistant *Staphylococcus*

Aureus (MRSA), *Escherichia Coli, Vibrio Cholera* and *Streptococcus Pneumoniae. Eur. Rev. Med. Pharmacol. Sci.* 2012, *3*, 105–111.

14. Gupta, S.; Gupta, S. M.; Kumar, N. Role of Chlorophyllase in Chlorophyll Homeostatis and Post-Harvest Breakdown in *P. Betle* L. Leaf. *Ind. J. Biochem. Biophys.* **2011,** *48*, 353–360.

15. Hara-Kudo, Y.; Kobayashi, A.; Sugita-Konishi, Y.; Kondo, K. Antibacterial Activity of Plants Used in Cooking for Aroma and Taste. *J. Food Prot.* **2004,** *67*, 2820–2824.

16. Holder, I. A.; Boyce, S. T. Agar Well Diffusion Assay Testing of Bacterial Susceptibility to Various Antimicrobials in Concentrations Non-Toxic for Human Cells in Culture. *Burns* **1994,** *20*, 426–429.

17. Hortensteiner, S. Chlorophyll Degradation During Senescence. *Ann. Rev. Plant Biol.* **2006,** *57*, 55–57.

18. Hutadilok, N. T.; Chaiyamutti, P.; Panthong, K.; Mahabusarakam, W.; Rukachaisirikul, V. Antioxidant and Free Radical Scavenging Activities of Some Plants Used in Thai Folk Medicine. *Pharmaceut. Biol.* **2006,** *44*, 221–228.

19. Khan, J. A.; Kumar, N. Evaluation of Anti-Bacterial Properties of Extracts of *Piper Betle* Leaf. *J. Pharmaceut. Biomed. Sci.* **2011,** *11*, 1–3.

20. Kumar, N.; Gupta, S.; Tripathi, A. N. Gender-Specific Responses of *Piper Betle* L. to Low Temperature Stress: Changes in Chlorophyllase Activity. *Biologia. Plantarum* **2006,** *50*, 705–708.

21. Kumar, N.; Mishra, P.; Dube, A.; Bhattacharya, S.; Dikshit, M. *Piper Betle* Linn. A Maligned Pan-Asiatic Plant with an Array of Pharmacological Activities and Prospects for Drug Discovery. *Curr. Sci.* **2010,** *99*, 922–932.

22. Lin, W. S.; Song, X. Z. Clinical and Experimental Research on a Kidney Toxifying Prescription in Preventing and Treating Children's Hearing Loss Induced by Amino-Glycoside Antibiotic Toxicity. *Chin. J. Mod. Develop. Trad. Med.* **1989,** *9*, 402–404.

23. Malini, T.; Arunakaran, J.; Aruldhas, M. M.; Govindarajulu, P. Effects of Piperine on the Medical Activities of *Piper* Species. *J. Med. Aromat. Plant Sci.* **1999,** *18*, 302–321.

24. Medeiros, E. S.; Gimaraes, E. F. Piperaceae do Parque Estadual do Ibitipoca, Minas Gerais, Brasil (Piperaceae of Ibitipoca State Park, Minas Gerais, Brazil). *Boletim de Botânica da Universidade de São Paulo* (Botany Bulletin of the University of São Paulo) **2007,** *25*, 227–252.

25. Mitchell, A. M.; Mitchell, T. J. *Streptococcus Pneumoniae:* Virulence Factors and Variation. *Clin. Microbiol. Infect.* **2010,** *16*, 411–418.

26. Moran, J. A.; Mitchell, A. K.; Goodmanson, G.; Stockburger, K. A. Differentiation among Effects of Nitrogen Fertilization Treatments on Conifer Seedlings by Foliar Reflectance: A Comparison of Methods. *Tree Physiol.* **2000,** *20*, 1113–1120.

27. Munurazzaman, M.; Chowdhury, M. B. R. Sensitivity of Fish Pathogenic Bacteria to Various Medicinal Herbs. *Bangla. J. Vet. Med.* **2004,** *2*, 75–82.

28. Nathalie, G.; Stefanie, D. B.; Ronan, S.; Yusuke, J.; Eunyoung, C. S. D.; Twiggy, V. D., Liesbeth, D. M.; Detlef, W.; Yuji, K.; Mark, S.; Gerrit, T. S. B.; Dirk, I. Increased Leaf Size: Different Means to an End. *Plant Physiol.* **2010,** *153*, 1261–1279.

29. Noga, E. J. Fish Disease: Diagnosis and Treatment, 2nd ed.. In *Mosby-year Book*; Wiley-Blackwell Publishing Inc.: New York, 2010; p 328.

30. Pandey, B.; Khan, S. Comparative Study of Antimicrobial Activity of Indian Spices. *Ind. J. Life Sci.* **2013**, *3*, 1–6.

31. Parmer, V. S.; Jain, S. C.; Bisht, K. S.; Jain, R.; Taneja, P.; Jha, A.; Tyagi, O. P.; Prasad, A. K.; Weengel, J.; Olsen, C. E.; Boll, M. Phytochemistry of Genus *Piper. Phytochem.* **1997**, *46*, 597–673.

32. Parthasarathy, U.; Asish, G. R.; Zachariah, T. Z.; Saji, K. V.; Johnson, K. G.; Jayarajan, J.; Mathew, P. A. Spatial Influence on the Important Biochemical Properties of *Piper Nigrum* Linn Leaves. *Nat. Prod. Rad.* **2008**, *7*, 444–447.

33. Penuelas, J.; Filella, I. Visible and Near-Infrared Reflectance Techniques for Diagnosing Plant Physiological Status. *Trend. Plant Sci.* **1998**, *3*, 151–156.

34. Praveen, S.; Mishra, S.; Hemant, S. Therapeutic Role of Indian Spices in the Treatment of Gastrointestinal Disease Caused by *Vibrio* Species. *Int. J. Innov. Res. Sci. Engineer. Technol.* **2013**, *2*, 2371–2375.

35. Rajeshbabu, P.; Rasool, S. K.; Sheriff, M. A.; Sekar, T. *In Vitro* Antibacterial Activity of *Piper Betle* L. against *Staphylococcus Aureus* and *Streptococcus Pneumoniae. J. Pharm. Res.* **2011**, *4*, 2223–2225.

36. Roberts, R. J. *Fish Pathology*, 4th ed.; Blackwell Publishing Ltd.: London, 2012; p 243.

37. Sadasivam, S.; Manickam, A. *Biochemical Methods,* Third ed.; New Age International (P) Ltd. Publishers: New Delhi, 2008; p 251.

38. Scott, I. A.; Jensen, H. R.; Philogene, B. J. R.; Arnoson, J. T. A. Review of *Piper* Spp. (Piperaceae) Phytochemistry, Insecticidal Activity and Modes of Action. *Phytochem. Rev.* **2008**, *7*, 65–75.

39. Sengupta, S.; Ghosh, S.; Bhattacharjee, S. *Allium* Vegetables in Cancer Prevention. *Asian Pac. J. Cancer Prevent.* **2004**, *5*, 237–245.

40. Singh, S.; Waman, A. A.; Bohra, P.; Gautam, R. K.; Dam, S. Conservation and Sustainable Utilization of Horticultural Biodiversity in Tropical Andaman and Nicobar Islands, India. *Genet. Res. Crop Evol.* **2016**, *63*, 1431–1445.

41. Sreekumar, P. V.; Ellis, J. L. Six Wild Relatives of Beetel Vine from Great Nicobar. *J. Andaman Sci. Assoc.* **1990**, *6*, 150–152.

42. Takenaka, A. Effects of Leaf Blade Narrowness and Petiole Length on the Light Capture Efficiency of a Shoot. *Ecol. Res.* **1994**, *9*, 109–114.

43. Tong, S. Y. C.; Davis, J. S.; Eichenberger, E.; Holland, T. L.; Fowler, V. G. *Staphylococcus Aureus* Infections: Epidemiology, Pathophysiology, Clinical Manifestations, and Management. *Clin. Microbiol. Rev.* **2015**, *28*, 603–661.

44. Tripathi, A. K.; Gautam, M. Biochemical Parameters of Plants as Indicators of Air Pollution. *J. Environ. Biol.* **2007**, *28*, 127–132.

45. Trivedi, M. N.; Khemani, A.; Vachhani, U. D.; Shah, C. P.; Santani, D. D. Pharmacognostic, Phytochemical Analysis and Anti-Microbial Activity of Two *Piper* Species. *Pharma. Glob.* **2011**, *2*, 1–4.

46. Tuntiwachwuttikul, P.; Phansa, P.; Pootaeng, O. Chemical Constituents of the Roots of *Piper Sarmentosum. Chem. Pharmaceut. Bull.* **2006**, *54*, 149–151.

47. Vaghasiya, Y.; Nair, R.; Chanda, S. Investigation of Some Piper Species for Antibacterial and Anti-Inflammatory Properties. *Int. J. Pharm.* **2007**, *3*, 400–405.

48. Waman, A. A.; Bohra, P. Sustainable Development of Medicinal and Aromatic Plants Sector in India: An Overview. *Sci. Cult.* **2016**, *82*, 245–250.

49. Waman, A. A.; Karanjalkar, G. R. Diversifying the Income Avenues through Herbal Spices. *Ind. J. Arecanut, Spices Med. Plants* **2010,** *12,* 18–21.

50. WHO. Traditional Medicine Growing Needs and Potential. In *WHO Policy Perspectives on Medicine Bulletin 002*; World Health Organization (WHO): Geneva, Switzerland, 2002; p 31.

51. Zahidi, S. A.; Aemeur, F.; Mousadik, A. E. Variability in Leaf Size and Shape in Three Natural Populations of *Argania Spinosa* (L.). *Int. J. Curr. Res. Acad. Rev.* **2013,** *1,* 13–25.

52. Zaidan, M. R. S.; Rain, N. A. *In Vitro* Screening of Five Local Medicinal Plants for Anti-Bacterial Activity Using Disc Diffusion Method. *Trop. Biomed.* **2005,** *22,* 165–170.

CHAPTER 6

ACONITE: ETHNOPHARMACOLOGICAL BENEFITS AND TOXICITY

YOGINI S. JAISWAL and LEONARD L. WILLIAMS

ABSTRACT

Aconite is reputed as queen of poisons and is revered for its anti-inflammatory properties in traditional systems of medicine (TSMs), despite its fatal toxic attributes. Ayurveda and Traditional Chinese Medicine (TCM) use different species of the *aconitum* genus for their therapeutic benefits and treatment of various ailments. Several species of the *aconitum* genus are listed as critically endangered species, by the International Union for Conservation of Nature (IUCN). The preparation methods for roots of *aconitum* play a critical role in altering the chemistry of alkaloids and toxicity effects of the prepared drug. Ayurveda and TCM have preparation methods that vary on the type of solvents and temperatures used for preparation of the processed drug from the raw roots. There are several case reports published till date that emphasize on the toxicity of aconite caused due to inadequate or improper processing of the drug. Despite the toxicity, aconite forms an important pillar of TSMs for its excellent therapeutic activity in treatment of rheumatalgia, severe generalized pain, menorrhagia, and cold. The present chapter focuses on the overview of ethnopharmacology, toxicity, chemistry, as well as applications and formulations of aconite in various TSMs.

6.1 INTRODUCTION

Aconite belongs to large genus, which consists of about 300 species distributed worldwide.[72] It belongs to the genus *Aconitum* and Ranunculaceae family, and most of its species are grown in higher altitudes. Despite their toxicities, aconite species are used widely in traditional systems of medicines (TSMs) for their analgesic, antineuralgic, antirheumatic, diuretic, anthelmintic, and antipyretic benefits.[51,57] The natural habitats of aconite species have been exploited for their uses in TSMs, and this has led to some of the species facing the danger of extinction. *Aconitum heterophyllum* Wall. Ex Royle has been listed as a critically endangered species by the International Union for Conservation of Nature and Natural Resources (IUCN).[82]

The propagation in aconite species occurs through seeds. Researchers have taken measures to emphasize the importance of aconite species and its conservation by publication of reports that detail the in vitro seed germination and growth conditions for propagation of its species.[74] The vegetative propagation of the seeds of aconite has issues of slow growth and seed dormancy (characteristic of Ranunculaceae family) that leads to impairment of its morphological growth.[6,64]

In this chapter, three species of aconite have been discussed: *A. kusnezoffii* (Caowu) and *A. carmichaelii* (Chuanwu) are used in Traditional Chinese Medicine (TCM), and *A. heterophyllum* Wall. Ex Royle (Atis) is used in one of the TSMs of India called as Ayurveda.[12,18,34,36] *A. kusnezoffii*, *A. carmichaelii* and *A. heterophyllum* and their formulations are recorded as traditional medicines in Chinese Pharmacopoeia and Ayurvedic formulary of India, respectively for the treatment of ailments mentioned in this chapter.[13,17]

> *A. heterophyllum* Wall. Ex Royle is commonly called as "*Atis*" and is used as an anti-pyretic, anti-inflammatory, anthelmintic, and diuretic drug in Ayurveda. It is found in various colors including black, red, white, and yellow of which the white variety is most commonly used. Its habitats are found in Nepal, Sikkim, and Himalayas at an altitude of up to 5000 m.[66,81] Homeopathy (one of the allied systems of medicine in India) also suggests the use of *A. heterophyllum* as an analgesic and nerve sedative.[80]

Aconite is considered as an important pillar of the legacy of TCM. *A. kusnezoffii* and *A. carmichaelii* are the two most widely used species, grown in the northern and south-west parts of China.[34] The growth areas include provinces of Hunan, Sichuan and Hubei, and certain regions around Jiangyou. In TCM, aconite is used for the treatment of diarrhea, colds, arthralgia, cardiac problems, and edema.[17]

This chapter discusses beneficial therapeutic effects of aconite, while also highlights the toxicities and fatal cases that have been reported due to improper dosage, processing, and formulation of this drug. This information may be helpful to patients, herbal drug industrialists, and traditional medicine practitioners in avoiding fatal toxicities caused due to improper use of the drug.

6.2 HISTORY

Aconite species have been of high value in both TCM and Ayurveda for many centuries.[28] In ancient TCM, the two species *A. kusnezoffii* and *A. carmichaelii* were evaluated for their quality and therapeutic efficacy based on their morphological characteristics. In the *Compendium of Materia Medica*, aconite roots with a wide base and few longitudinal projections are described as roots with good efficacy.[96] In the *Origins of the Materia Media*, roots with more than six to eight projections are considered to be of higher quality.

The records for use of aconite species have been found in the *Atharva Veda*, which is revered as one of the most ancient records of Ayurvedic Materia Medica.[45,58] Literature reveals the use of aconite paste as an arrow poison in warfare in ancient India. In *Sushruta Samhita* (ancient text of Ayurveda dated ca. 300 C.E.), the *A. ferox* species is referred to as "*vatsanabha.*" Aconite is also described in *Sharangadhara Samhita* in 1363 AD and Chakra Dutta in 1050 AD for its properties in treating diarrhea, congestion, cough, and male infertility. The *A. ferox* species were found by European travelers in their journey to Nepal during the 19th century and it gained an upsurge in trade to Ladakhh during this era.[68]

The word "Aconite" is coined from the Greek word "*Akoniton,*" which was used by the Theophrastus to describe a poisonous plant called "hemlock." Aconite has been historically mentioned in TSMs of various countries and it gained its importance for the diterpene alkaloidal content and numerous species within its genus that range from being nonpoisonous to deadly poisonous.[68]

6.3 PHYTOCHEMICAL COMPOSITION

The chemical structures of aconitum alkaloids are shown in Figure 6.1. Based on the various substitution groups at the C_8 and C_{14} positions in the chemical structure of alkaloids, the toxic alkaloids in aconite species can be classified into three groups: [25,42,71,91]

A. Diester diterpene alkaloids (DDAs): The DDAs comprise of highly toxic alkaloids, which include hypaconitine, aconitine, and mesaconitine. The highly toxic aconitine is referred to as "queen of poisons."

B. Monoester diterpene alkaloids (MDAs): The MDAs comprise of
    low toxic constituents that include atisine, delphatine, deltaline, and
    benzoylaconine. The MDAs are devoid of an acetyl group at the C_8
    position in their chemical structure.
C. De-esterified diterpenoid alkaloids (DEDAs): The DEDAs are
    nontoxic compounds, and they are devoid of an acetyl and benzoyl
    groups at the C_8 and C_{14} positions. Examples of such compounds
    include lycoctonine and karakoline.

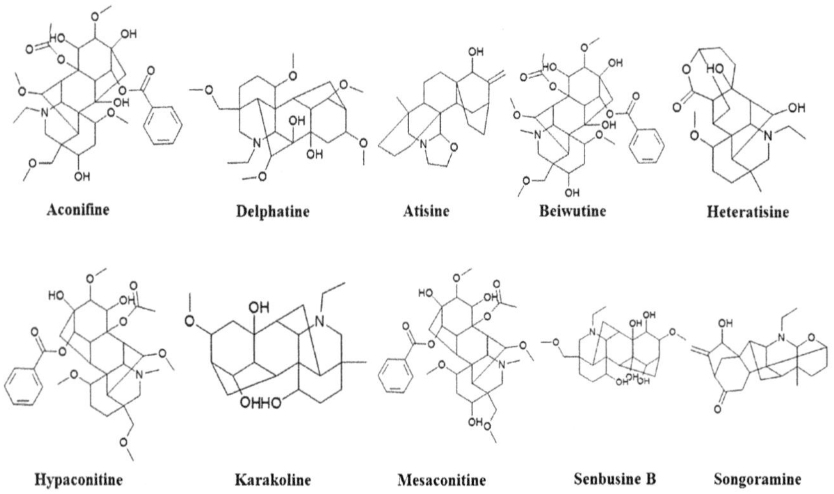

| Aconifine | Delphatine | Atisine | Beiwutine | Heteratisine |

| Hypaconitine | Karakoline | Mesaconitine | Senbusine B | Songoramine |

FIGURE 6.1   Alkaloidal compounds found in aconite species.

Due to the presence of both an acetyl and a benzoyl group at C_8 and C_{14}
positions, the DDAs are not very stable during a hydrolysis reaction. There
is a stepwise loss of acetyl and benzoyl groups, which lead to formation of
MDAs and DEDAs in the respective steps.[51] An illustration of this two-step
hydrolysis is shown in Figure 6.2.

6.4   PHARMACOLOGICAL/THERAPEUTIC EFFECTS

In Ayurveda and TCM, aconite is commonly used for cold and cough and
pain-related conditions. Despite its toxic effects, it is widely used in the TSM
with great precaution about the processing method, the source of the plant
(the species used), the dose, other pathological issues of the patient, and the
herbs that are used in combination with aconite. The use of aconite during

pregnancy is cautioned. The herbs incompatible with aconite are: *Fructustrichosanthis, Bulbusfritilliariae, Radix Ampelopsis*, and *Rhizomabletillae*. In Ayurveda, herbs are evaluated for their pharmacological effects based on their physicochemical properties such as: taste (*Rasa*), physiological effect (*Virya*), any unusual or adverse reactions of the drug (*Prabhava*), after taste of the drug (*Vipaka*), and morphological properties (*Guna*).[62,97] The pharmacological effects of aconite based on the philosophy of Ayurveda are detailed in Table 6.1.

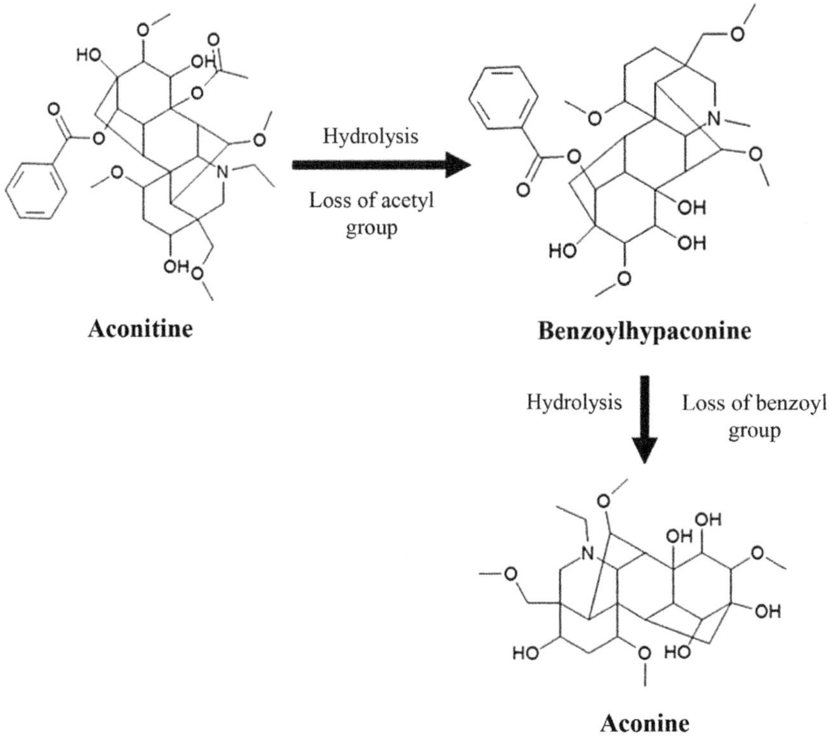

FIGURE 6.2   Hydrolytic conversion of alkaloids in aconite.

In TCM, it is believed that the processed form of aconite called "Fuzi" can have variations in its potency depending on the location of collection of roots, season of collection, processing method used, and mode of administration. Based on the philosophy of TCM, Fuzi is believed to reinforce *Yang* (principle of body associated with heat). Fuzi is believed to treat impotence, nausea, cold limbs, and diarrhea.[97] Reports on polyherbal formulation of Fuzi indicate that Fuzi is also used to treat coronary heart disorders, myocardial infarction, and blood pressure issues.[96] The Korean Pharmacopoeia describes processed aconite to be effective in treating neuralgia and rheumatoid arthritis. In Japanese pharmacopoeia, the properties of processed aconite roots (called "*Bushi*") are described. With the discovery of better analgesics and drugs to treat conditions that aconite was used for, aconite is no longer used in Western medicine.[94] Several studies have been performed and reported for in vivo and in vitro models, to study the pharmacological effects on aconite and its processed products. Selected activities are described in this section.

TABLE 6.1 Physicochemical and Pharmacological Properties of Aconite Species Based on the Philosophy of Ayurveda.

Physicochemical properties	Pharmacological effects
Virya	It improves digestion by increasing *pitta* (humor that regulates hunger, thirst, and body temperature).
	It causes a reduction in *kapha* (humor that regulates lubrication of joints) and promotes prokinesis.
	It produces an antiinflammatory effect by reduction of *vata* (humor that regulates electrolyte balance, and elimination of waste products from the body).
Rasa	Reduces fever by reduction of *kapha*.
	Promotes prokinesis by reducing *kapha*.
	Causes a reduction in the production of mucous by reducing *kapha*.
Guna	Causes a reduction in inflammation and edemas by reducing *kapha*.
	Causes a reduction in the production of mucous and mucolysis by reducing *kapha*.

6.5   ACTIVITY ON CARDIOVASCULAR SYSTEM

Fuzi products of TCM have been tested by several researchers to evaluate its effect as a cardiotonic in various in vivo models. Effect of Fuzi was tested on bull-frog hearts, hearts of guinea pigs, cats, and rabbits in independent studies.[53,71,87] The cardiotonic effect is believed to be due to excitement of beta receptors and release of catecholamine. The constituents from the roots (such as: mesaconine, salsolinol, senbusine A, neoline, and hetisine) are reported to cause the cardiotonic effect.[15,16,27,33,35,41,70] The polysaccharides present in Fuzi are reported to have protective effect on the cardiomyocytes through following four mechanisms:

a) Reduction of mitochondrial injury by maintaining calcium ion concentration[20];
b) Free radical scavenging effect[55];
c) Enhancing signal transduction and protection of mitochondria to inhibit apoptotic events[57]; and
d) Increasing the gene expression of superoxide dismutase.[56]

Fuzi is also reported to have antihypertensive, antithrombotic, and anti-arrhythmic effects.[54,89] However, some studies also caution the toxic arrhythmogenic effects of aconitine.[50,97]

6.5.1   ANTIINFLAMMATORY AND ANALGESIC ACTIVITIES

Fuzi is reported to exhibit analgesic effect by mediating antinociceptive action through opioid receptors and reduction in cytokine production. Certain reports also indicated an anti-rhinitis effect in in vivo allergy models.[47,85,93] The anti-inflammatory effect of extracts of Fuzi is reported to be due to inhibition of interleukin-1ß and nitric oxide.[51,88] The ethanol extracts of *A. heterophyllum* processed by *shophahara* method of Ayurveda is reported to exhibit antinflammatory activity.[51] In Ayurveda, the dry powder and decoction of *A. heterophyllum* are used as antipyretics.[84]

6.5.2   ANTICANCER ACTIVITY

The anticancer effect of Fuzi has been evaluated by several researchers in various models, which include breast cancer, hepatocellular, and lung carcinoma.[3,45] The mechanisms postulated for this activity are[64,78]: (1) Induction of apoptosis of the tumor cells; (2) Increase in immunological responses; and (3) Regulation of oncogens.

6.5.3   IMMUNOMODULATORY ACTIVITY

The use of Fuzi in moxibustion (TCM treatment by burning the herb on certain tissue points of the body) is reported to enhance immunity in elderly patients.[24] Chen et al. have reported the immunomodulatory effect of Fuzi due to enhancement of the production of interleukins.[48] Atal et al.[16] tested the roots of *A. heterophyllum* for their immunomodulatory effect on sheep red blood cells. They indicated that *A. heterophyllum* reduces the humoral component of the immune system.

6.5.4   ANTIHYPERLIPIDEMIC AND ANTIDIABETIC ACTIVITIES

The hypoglycemic effect of polysaccharides from Fuzi was evaluated by researchers in alloxan-induced and streptozotocin-induced rat models.[4,42] The hypoglycemic effect of the polysaccharides was reported to be due to increase in the activity of hepatic phosphofructokinase and glucose utilization rate of adipocytes.[29,91] *A. heterophyllum*is is reported to lower cholesterol, triglycerides, low-density lipoprotein cholesterol (LDL-c), and high-density lipoprotein cholesterol (HDL-c).[30] Fuzi is reported to exhibit antihyperlipidemic activity by[29,91]:

   a)   Reducing the protein expression of hepatic low-density receptors of lipoproteins;
   b)   Down regulation of expression of hepatic methylglutaryl coenzyme reductases; and
   c)   Upregulating of mRNA expression.

6.5.5   ANTIBACTERIAL ACTIVITY

Shaheen et al. isolated phytoconstituents from roots of *A. heterophyllum* and evaluated their antibacterial effect against *Salmonella typhi*, *Escherichia coli*, *Pseudomonas aeruginosa*, and *Shigellaflexineri*. The compounds found to show antibacterial activity were: 13-hydroxylappaconitine, delphatine, lappaconitine, and lycoctonine.[32]

6.6   TOXICITY

The history of toxicity of aconite dates back to the ancient Greek period. Greek mythology explains aconite poison to have developed from the saliva of Cerberus (a three-headed dog) with whom Hercules had an encounter.[75] Historical writings also mention that the shape of aconite flower is the shape of the cup that was prepared by Medea for Theseus. Roman history records mention that Nero poisoned Claudius with aconite to obtain the throne, after which the growth of this plant was banned in the Roman Empire. Shakespeare in his novel *Romeo and Juliet* mentioned aconite as the poison that was consumed by Romeo for his suicide.[1] In ancient writings of TCM, *Shennong Bencao Jing* describes the juice of aconite—*shezou* to be the juice that would kill animals. TCM alchemist *Tao Hongjing* describes aconite as a poison that when put on an arrow head can kill a beast within ten steps after being shot with the poisoned arrow.[58] In 1930s, aconite preparations were used in the United States of America and Canada for treating rheumatism, as an analgesic and painkiller. Reports of toxicities led to the ban of this drug.[78,79]

Several researchers indicate the fatal events due to aconite through following mechanisms[21,76]: apoptosis of vital organs such as heart and liver; changes in release of neurotransmitters; and changes in voltage-gated sodium ion channels. Due to these actions, aconite affects the central nervous system, heart and liver and can cause cerebral paralysis, respiratory failure, hypotension, paralysis of limbs, and cardiac arrest. There have been no successful treatments identified for aconite poisoning and only supportive treatments are provided in such cases. The symptoms reported for aconite poisoning that can aid in diagnosis are: excessive salivation, increased palpitation, dizziness, arrhythmia, shock, and coma.

Hundreds of deaths due to aconite poisoning have been reported till date.[5,13,19,22,43,89] The deaths due to aconite poisoning can be avoided by proper education of traditional medicine practitioners and patients about the appropriate mode and processing methods, root collection areas, its incompatibilities and doses, physiological status, and preexisting disease conditions of the patient.[9,48,60,94] The combination of aconite species is contraindicated[97] with: *Rhizomabletillae, Fructustrichosanthis, Bulbusfritillariae, Radixampelopsis* and *Rhizomapinelliae, Dhaturastramonium, and Ocimum sanctum.*[97]

6.7 PROCESSING METHODS

The processing of aconite may reduce or completely remove its toxicity. Researchers have identified the chemistry involved in toxicity potential of the herb. They have proved that the conversion of DDAs to MDAs and DEDAs leads to this change during processing.[25,35] The conversion efficiency and the resultant reduction in toxicity is highly dependent on the process, the solvent or media used for processing, the duration of processing, and the geographical source of the roots.

Paozhi (processing method) of herbs is an important aspect of TCM. For *aconitum*, various processing methods have been developed during the rule of different dynasties in the history of China.[30] An overview of few methods used for processing of aconite is shown in Table 6.2. The processing of Chuanwu with wine has been documented in ancient literature of TCM. The clinical dose of Fuzi is suggested to be 3–15 g per day.[11] The processing methods of aconite in TCM can be broadly classified as:

a) Soaking (treating herb with solvents such as water, wine, honey, etc.);

b) Heat/ fire treatment (stir frying, roasting, or baking); and

c) Combination of solvent and heat treatment (decoction or steaming).

In Ayurvedic pharmacy (*Rasa shastra*), processing of drugs (*Samskaras*) is divided into steps, such as[7,11,25]: (1) purification or detoxification *(Shodhana)* and(2) formulation *(Bhaishajyakalpana)*.

In Ayurveda, three methods for processing of aconite involve use of three different solvents: cow urine, cow milk, and goat milk.[8,59,61,66] The processing method using cow urine is considered as the most superior method of the three methods. It is demonstrated in scientific reports that the acidic pH of cow urine (pH = 6.9) and exposure to heat can lead to conversion of DDAs to MDAs and DEDAs. Reports also indicate that cow urine enhances the bioavailability of drugs and helps in detoxification.[10,23,33,38]

The process of treatment of aconite with cow urine involves dipping the aconite root pieces in cow urine in an earthen or stone vessel and keeping it in sunlight for three days, with change of cow urine from the vessel, on each day. For treatment with cow milk, the aconite root pieces are packed in meshed cloth and suspended in a pot (*Dolayantra*) with boiling cow milk. The boiling is carried out for three to six hours. The third method involves use of goat milk instead of cow milk and boiling the aconite roots in the same way as mentioned for the second method with cow milk.[37]

6.8   FORMULATIONS AND PRODUCTS

There are numerous formulations in Ayurveda and TCM in the form of decoctions, powders, and polyherbal preparation that use aconite after processing and are used with caution by traditional medicine practitioners. Caution must be taken with the mode of administration and dose even if the drugs are appropriately processed. A list of formulations that use aconite in TCM and Ayurveda practice are listed in Tables 6.3 and 6.4, respectively.[63,67,86,97]

TABLE 6.2   Ancient Processing Methods of Aconite in TCM.

Name of the preparation	Name of the dynasty when the processing method originated	Processing method
Caowu	Song dynasty	Charring the roots, stir frying until the color of roots darken
	Jin Yuan dynasty	Soaking in water, charring, stir frying with salt
	Qing dynasty	Stir frying with moistened paper, decoction with mung soy beans
	Ming dynasty	Soaking in water, rice water, vinegar, ginger juice, and stir frying
Chuanwu	Song dynasty	Stir fry the roots with wine, sesame oil, lard, or salt. Soaking in rice water for four weeks, decoction with soy beans, soaking in salt solution
	Jin Yuan dynasty	Stir frying and charring
	Qing dynasty	Roasting, decoction with mung soy beans, wine and ginger juice, and baking
	Ming dynasty	Steaming with soy beans, soaking in water, rice water, vinegar, ginger juice, and stir frying
Fuzi	Jin dynasty	Dipping aconite in vinegar, charring and baking
	Tang dynasty	Soaking the herb in rice vinegar, heating the herb in ashes, and stir frying with yellow honey, decoction with ginger juice
	Han dynasty	Soaking the herb in water for one week with regularly changing of the water and peeling the roots, stir frying with honey
	Song dynasty	Soaking the herb in water for one week with regularly changing of the water, soaking in ginger juice, roasting in wet paper.
	Qing dynasty	Stir frying with salt, decoction with *Euphorbia kansui*, and soy bean juice.
	Ming dynasty	Decoction with ginger juice, *Coptidis rhizoma* and juice of Rehmanniae Radix.

6.9   RELATED SPECIES

There are several related species of aconite, which have phytoconstituents with structures closely related to aconitine. *Aconitum Lycoctonum* Linn. Is reported to contain acolyctine lycoctonine that is postulated to be identical to *pseudaconine* and aconine. *Aconitum palmatum* Don. It is found in Nepal and is called "Bish" or "Bikhma." The *Taxus* and *Delphinium* spp. Contains terpenoid alkaloids and are used as aconite in various countries. *A. spicatum*, *A. laciniatum*, *A. balfourii*, and *A. deinorrhizum* are other species of aconite imported from India and Pakistan. In Japan, *Aconitum japonicum*, which has good cardiotonic properties and is commonly used in traditional medicine.[39,73]

TABLE 6.3　Formulations of Aconite used in TCM.

Composition	Pharmacological activity/ therapeutic effect	Name of the formulation
Contain decocted herbs: Radix aconiti, Herbaepimedii, Radix angelicae sinensis, Cortex cinnamomi, Radix Paeoniae alba, and *Radix codonopsis*	Analgesic effect, joint pain reliever, and blood circulation enhancer	Fugui Gutong Pian (tablet): approved by the China Food and Drug Administration (CFDA)
Contains *Radix glycyrrhizae*, Fuzi (56%), and *Rhizoma zingiberis*	Used for treatment of mycocardial infarction and cold extremities	Sini Tang (Decoction)
Rhizoma zingiberis Recens, Fuzi (21%), *Rhizoma atractylodis Macrocephalae*, and *Radix paeoniae* Alba.	Used to treat polyuria, cardiac edema, and bronchitis	Zhenwu Tang (Decoction)
Fuzi (20%), *Radix glycyrrhizae, Rhizoma atractylodis Macrocephalae, Radix codonopsis*, and *Rhizoma zingiberis*	Used to relieve abdominal pain and coldness of limbs	Fuzi Lizhong Wan (tablet): approved by the CFDA
Fuzi (12%), *Rhizoma dioscoreae, Radix rehmanniae, Fructus corni, Ramulus cinnamomi, Poria, Rhizoma alismatis*,and *Cortex moutan*	Used in treatment of asthma, diarrhea, and nephritis	Jingui Shenqi Wan (tablet): approved by the CFDA

TABLE 6.4　Formulations of Aconite Used in Ayurveda.

Name of formulation	Pharmacological activity/therapeutic effect
Amavat vidhvansa rasa	Treatment of rheumatoid arthritis
	Treatment of *Vata*-related complications
	Used as a cardiotonic
Maha visgarbha-Taila	Used to treat rheumatoid arthritis and neuralgic pain
Sanji vanivati	Treatment of rheumatoid arthritis
Suta shekhara Rasa	Used to treat rheumatoid arthritis and neuralgic pain
Tribhu vanaKirti-Rasa	Used in treatment of typhoid

6.10 SUMMARY

Aconitine from aconite is reputed as the queen of poisons and aconite is revered for its antiinflammatory properties in TSMs, despite its fatal toxic attributes. Ayurveda and TCM use different species of the genus—aconitum, for their therapeutic benefits and treatment of various ailments. Ayurveda and TCM have preparation methods that vary in the type of solvents and temperatures used for preparation of the processed drug from the raw roots. Despite the toxicity, aconite forms an important pillar of TSMs for its excellent therapeutic activity in treatment of rheumatalgia, severe generalized pain, menorrhagia, and cold. The present chapter focuses on providing an overview of the ethnopharmacology, toxicity, chemistry, applications, and formulations of aconite in various TSMs.

KEYWORDS

- **aconite**
- **ethnopharmacology**
- **toxicity**
- **radix aconite**
- **Traditional Chinese Medicine**

REFERENCES

1. Ahmad, M.; Ahmad, W.; Ahmad, M.; Zeeshan, M.; Shaheen, O.; Shaheen, F. Norditerpenoid Alkaloids from the Roots of *Aconitum heterophyllum* with Antibacterial Activity. *J. Enzyme Inhib. Med. Chem.* **2008,** *23*, 1018–1022.
2. Anonymous. *The Ayurvedic formulary of India.* Ministry of Health and Family Welfare, Government of India, New Delhi, 1977; pp. 250–252.
3. Anonymous. *The Ayurvedic Pharmacopoeia of India*, Ministry of Health and Family Welfare, Department of Indian Systems of Medicine and Homoeopathy, Government of India, New Delhi, 2001; pp. 129–130.
4. Atal, C. K.; Sharma, M. L.; Kaul, A.; Khajuria, A. Immunomodulating Agents of Plant Origin. I: Preliminary Screening.*J. Ethnopharmacol.* **1986,** *18*, 133–141.
5. Bause, G. S. Monkshoods: Heart-Stopping Neurotoxins Behind Hanaoka's Mafutsusan. *Anesthesiology* **2015,** *123*, 376.
6. Beigh, S. Y.; Nawchoo, I. A.; Iqbal, M. Cultivation and Conservation of *Aconitum heterophyllum*: A Critically Endangered Medicinal Herb of the Northwest Himalayas. *J. Herbs, Spices Med. Plants* **2006,** *11*, 47–56.

7. Belge, R. S.; Belge, A. R. Ayurvedic Shodhana Treatments and Their Applied Aspect with Special Reference to Loha.*IOSR J. Pharm. Biol. Sci.* **2012,** *7,* 45–49.

8. Bhadauria, H. Cow Urine—A Magical Therapy.*Int. J. Cow Sci.* **2002,** *7,* 32–36.

9. Chan, T. Y. Incidence and Causes of *Aconitum* Alkaloid Poisoning in Hong Kong from 1989 to 2010. *Phytother. Res.* **2015,** *29,* 1107–1111.

10. Chauhan, R. S. *Panchagavya* Therapy (Cow pathy)—Current Status and Future Directions. *Indian Cow* **2004,** *2004,* 3–7.

11. Chen, C. X.; Xu, S. J. Progress in Material Basis of Radix *Aconite Lateralis Prepapata* Compatibility Radix Glycyrrhizae and *Rhizoma Zingiberis. Traditional Chinese Drug Res. Clin. Pharma.* **2006,** *17,* 472–476.

12. Chen, D. H.; Liang, X. T. Studies on Isolation and Structural Determination of Salsolinol, the Constituents of Fuzi—Part I. *ActaPharmacologicaSinica* **1982,** *17,* 792–974.

13. Chen, X.; Cao, Y.; Zhang, H.; Zhu, Z. Comparative Normal/Failing Rat Myocardium Cell Membrane Chromatographic Analysis System for Screening Specific Components that Counteract Doxorubicin-Induced Heart Failure from *Acontium carmichaelii. Anal. Chem.* **2014,** *86,*4748–4757.

14. Chen, X.; Wu, R.; Jin, H. Successful Rescue of a Patient with Acute *Aconitine* Poisoning Complicated by Polycystic Renal Hemorrhage. *Nippon Med. Sch.* **2015,** *82,* 257–261.

15. Chen, Y.; Chu, Y. L.; Chu, J. H. Alkaloids of the Chinese Drugs, Aconitum Spp.-IX Alkaloids from Chuan-wu and Fuzi *Aconitum carmichae* lidebx. *ActaPharmacologica Sinica* **1956,** *12,* 435–439.

16. Chen, Y. C. Preliminary Investigation on Mechanisms of Immunoregulatory Function of Radix inseng, radixa conitipraeparata and Shenfu Decoction. *Chinese Traditional Patent Medicine* **1994,** *16,* 30–31.

17. Chinese Pharmacopoeia Commission (CPC). *The Pharmacopoeia of the People's Republic of China.* China Medical Science Press: Beijing, China, 2010; pp. 177–178.

18. Chopra, R. N.; Chopra, I. C.; Handa, K. L.; Kapur, L. D. *Chopra's Indigenous Drugs of India.* U.N. Duhr and Sons Pvt. Ltd.: Calcutta, 1958; pp. 54–57.

19. Coulson, J. M.; Caparrotta, T. M.; Thompson, J. P. The Management of *Chinese Traditional Patent Medicine* Ventricular Dysrhythmia in Aconite Poisoning. *Clin. Toxicol.(Phila)* **2017,** *55,* 313–321. E-article: doi: 10.1080/15563650.2017.1291944.

20. Dang, W. T.; Miao, W. N.; Yang, X. F.; Jiang, C. Research on the Targeting of Calcineurin (CaN) on the Procedure When Fuzinoside Regulate Heart Failure. *Pharmacy and Clinics of Chinese Materia Medica* **2011,** *27,* 59–61.

21. El-Shazly, M.; Tai, C. J.; Wu, T. Y.; Csupor, D.; Hohmann, J.; Chang, F.R.; Wu, Y.C. *Facts on File.* Infobase Publishing: New York, 2010; pp. 154–155.

22. Feng, L; Wang, X. Q.; Fu, M. Strategy for Isolating Differentiation-Inducing Complementary DNAs from Human Esophageal Cancer Cell Line Treated with Retinoic Acid. *Sci China B.* **1992,** *35* (4), 445–454. PMID-1590919.

23. Fu, M.; Wu, M.; Qiao, Y.; Wang, Z. Toxicological Mechanisms of *Aconitum* Alkaloids. *Die Pharmazie* **2006,** *61,* 735–741.

24. Ganaie, J. A.; Srivastava, V. K. Effects of Gonadotropin Releasing Hormone Conjugate Immunization and Bioenhancing Role of Kamdhenu Ark on Estrous Cycle, Serum Estradiol and Progesterone Levels in Female Mus musculus. *Iranian J.Reproduct. Med.* **2010,** *7,* 70–75.

25. Gao, L. L.; Zeng, S. P.; Pan, L. T. Induction of differentiation of dendritic cells derived from hepatocellular carcinoma by Fuzi polysaccharides. *Chinese J. Clin. Oncol.* **2012,** *29,* 882–885.

26. Gu, H. Research on toxicity of clinical application of Fuzi. *Yunnan J. Trad. Chinese Med. Materia Medica* **2012**, *33*, 82–83.

27. Gutser, U. T.; Friese, J.; Heubach, J. F.; Matthiesen, T.; Selve, N.; Wilffert, B.; Gleitz, J. Mode of Antinociceptive and Toxic Action of Alkaloids of *Aconitum* Spec. *N-SArch. Pharmacol.* **1998**, *357*, 39–48.

28. Han, G. Y.; Liang, H. Q.; Zhang, W. D.; Chen, H. S.; Wang, J. Z.; Liao, Y. Z.; Liu, M. Z.; Shen, Q. H. Studies on the Alkaloids and a New Cardiac Principle Isolated from Jiangyou Fuzi. *Nat. Prod. Res. Develop.* **1997**, *9*,30–33.

29. Heiner, F. The Flagship Remedy of Chinese Medicine: Reflections on the Toxicity and Safety of Aconite.*Am. J. Chin. Med.* **2012**, *7*, 36–41.

30. Hikino, H.; Kobayashi, M.; Suzuki, Y.; Konno, C. Mechanisms of Hypoglycemic Activity of Aconitan-A, A Glycan from *Aconitum carmichaelii* Roots. *J. Ethnopharmacol.* **1989**, *25*, 295–304.

31. Hou, X. J.; Zhang, P.; Ma, F.; Tang, J. M. Reflection on Safety of Clinical Application of Fuzi. *Beijing J. Tradit. Chin. Med.* **2011**, *30*, 218–221.

32. Huang, L. Z.; Zhang, D. Y.; Wang, C. Y. Synthesis of Higenaming. *Acta Pharmaceutica Sinica* **1981**, *16*, 931–933.

33. Huang, X.; Tang, J.; Zhou, Q.; Lu, H.; Wu, Y.; Wu, W. Polysaccharide from Fuzi (FPS) Prevents Hypercholesterolemia in Rats. *Lipids Health Dis.* **2010**, *9*,9–12.

34. Jain, N. K.; Gupta, V. B.; Garg, R.; Silawat, N. Efficacy of Cow Urine Therapy on Various Cancer Patients in Mansour District, India—Survey.*Int. J. Green Pharm.* **2010**, *7*, 29–35.

35. Jaiswal, Y.; Liang, Z.; Ho, A.; Wong, L.; Yong, P.; Chen, H.; Zhao, Z. Distribution of Toxic Alkaloids in Tissues from Three Herbal Medicine Aconitum Species Using Laser Micro-Dissection, UHPLC-QTOF MS and LC-MS/MS Techniques. *Phytochemistry* **2014**, *107*, 155–174.

36. Jaiswal, Y.; Liang, Z.; Yong, P.; Chen, H.; Zhao, Z. A Comparative Study on the Traditional Indian Shodhana and Chinese Processing Methods for Aconite Roots by Characterization and Determination of the Major Components. *Chem.Central J.* **2013**, *7*, 169–171.

37. Judith, S.; Ming, Z.; Sonja, P.; Brigitte, K. Aconitum in Traditional Chinese Medicine– A Valuable Drug or An Unpredictable Risk? *J. Ethnopharmacol.* **2009**, *126*,18–30.

38. Khan, A.; Srivastava, V. Antitoxic and Bioenhancing Role of Kamdhenu Ark (Cow Urine Distillate) on Fertility Rate of Male Mice (*mus musculus*) Affected by Cadmium Chloride Toxicity. *Int. J. Cow Sci.* **2005**, *7*, 43–46.

39. Khanuja, S. P.; Kumar, S.; Shasany, A. K.; Arya, J.S. Use of Bioactive Fraction from Cow Urine Distillate ('*go-mutra*') as a Bio-Enhancer of Anti-Infective, Anti-Cancer Agents and Nutrients.U.S. Patent 6896907; Washington, DC: US Government; May 24, 2005, p. 21.

40. King, J.; Felter, H. W.; Loyd, J. U. *King's American Dispensatory*. Cincinnati, USA: Dispensatories, Eclectic - Ohio Valley Co., 1905; pp. 101–106.

41. Konno, C.; Murayama, M.; Sugiyama, K.; Arai, M.; Murakami, M.; Takahashi M.; Hikino, H. Isolation and Hypoglycemic Activity of Aconitans A–D, Glycans of *Aconitum carmichaelii*Roots. *Planta Medica* **1985**, *2*, 160–161.

42. Konno, C.; Shirasaki, M.; Hikino, H. Cardioactive principle of *Aconitum carmichaelii* roots. *Planta Medica*, **1979**, *35*, 150–155.

43. Lei, J.; Luo, Y. J.; Bian, Q. Q.; Wang, X. Q. Talatisamine: C(19)-diterpenoid Alkaloid from Chinese Traditional Herbal Chuanwu. *Acta Crystallographica:Section-EStructure Reports Online*, 2011, *E67*, O3145–O3146.

44. Li, H.; Liu, L.; Zhu, S.; Liu, Q. Case Reports of Aconite Poisoning in Mainland China from 2004 to 2015: A Retrospective Analysis. *J. Forensic Leg. Med.* **2016**, *42*, 68–73.

45. Li, H. R.; Zhao, B.; Du, G. J.; Liu, W. J.; Li, J. H.; Wang, Y. Y. Preliminary Study on the Anti-Tumor Effect of Fuzi Medicinal Liquor. *J. Henan Univ. (Med. Sci.)* **2013**,*32*, 84–87.

46. Li, Z. L. *Origins of the Materia Medica (Ben Cao Yuan Shi)*. People's Medical Publishing House: Beijing, China, 2007; p. 108.

47. Liang, S. Y.; Tan, X. M.; Gao, J.; Hu, Y. L.; Wang, W. F. Study on Acute Toxicity of Total Alkaloids of Processed Radix *Aconiti Lateralis* and Its Effects on Blood Histamine Contents and Pathomorphological Changes in Nasal Mucosa in Allergic Rhinitis Guinea Pigs. *China J. Tradit. Chin. Med. Pharma.* **2011**, *26*, 2986–2989.

48. Liao, Y. X.; Li, J. H. The Effect of Monkshood-Cake Moxibustion on the Red-Cell Immune Function of the Aged. *J. Hunan Coll. Tradit. Chin. Med.***1996**, *16*, 42–46.

49. Lin, C. C.; Chan, T. Y. K.; Deng, J. F. Clinical Features and Management of Herb-Induced Aconitine Poisoning. *Ann. Emerg. Med.* **2004**, *43*,574–579.

50. Lin, J. S.; Chan, C. Y.; Yang, C.; Wang, Y. H.; Chiou, H. Y.; Su, Y.C. Zhi - Fuzi: A Cardiotonic Chinese Herb—A New Medical Treatment Choice for Portal Hypertension. *Exp. Biol. Med.* **2007**, *232*, 557–564.

51. Liu, J. L.; Li, B. L. Experimental Treatment of Rheumatoid Arthritis by *Aconite lateralis* Radix Praeparata. *Chin. J. Exp. Tradit. Med. Formulae* **2011**, *17*, 184–187.

52. Liu, S.; Li, F.; Li, Y.; Li, W.; Xu, J.; Du, H. Review of Traditional and Current Methods to Potentially Reduce Toxicity of *Aconitum* Roots in Traditional Chinese Medicine. *J. Ethnopharmacol.* **2017**, *207*, 237–250.

53. Liu, X. X.; Jian, X. X.; Cai, X. F. Cardioactive C_{19}Diterpenoid Alkaloids from the Lateral Roots of *Aconitum carmichaeli*, Fuzi. *Chem. Pharm. Bull.* **2012**, *60*,144–149.

54. Liu, Y.; Ji, C. Effects of Fuzi Polysaccharide Post Conditioning on Expression of Manganese Superoxide Dismutase in Neonatal Rat Cardiomyocytes with Hypoxia-Reoxygenation. *Pharm. Clin. Chin. Materia Medica* **2011**, *27*, 53–56.

55. Liu, Y.; Ji, C. Protection of Fuzi Polymccharide Against Hypoxia-Reoxygenation Injury in Neonatal Rat Cardiomycytes and Its Mechanism. *Tradit. Chin. Drug Res. Clin. Pharmacol.* **2012**, *23*, 504–507.

56. Liu, Y.; Ji, C.; Wu, W. K. Effect of STAT3 in Mechanism of Fuzi Polysaccharides Post Conditioning Protecting Cardiomyocytes with Hypoxia-Reoxygenation in Neonate Rats. *J. Beijing Uni. Tradit. Chin. Med.* **2012**, *35*, 169–173.

57. Liu, Y.; Ji, C.; Wu, W. K. Metallothionein Mediates Protection by Fuzi Polymccharide on Neonatal Rat Cardiomyocytes with Hypoxia-Reoxygenation. *Chin. J. Exp. Tradit. Med. Formulae* **2012**, *18*, 172–175.

58. Nagarajan, M.; Gina, R.; Kuruvilla, K.; Kumar, S.; Venkatasubramanian, P. Pharmacology of Ativisha, Musta and Their Substitutes. *J. Ayurveda Integ. Med.* **2015**, *6*, 121–133.

59. Prasad, P. V. Atharvaveda and Its Materia Medica. *Bull. Indian Inst. Hist. Med. Hyderabad* **2000**, *30*, 83–92.

60. Prashith, T. R.; Nishanth, B. C.; Praveen, S. V.; Kamal, D.; Sandeep, M.; Megharaj, H. K. Cow Urine Concentrate: Potent Agent with Antimicrobial and Anthelmintic Activity.*J.Pharm. Res.* **2010**, *7*, 1025–1027.

61. Qiu, Y. H.; Yang, M. X. Elementary Analysis on Safety and Countermeasures of Clinical Excessive Application of Fuzi in Urumqi City Hospital of Chinese Traditional Medicine. *Xinjiang J. Tradit. Chin. Med.* **2012**, *30*, 56–58.

62. Randhawa, G. K. Cow Urine Distillate as Bioenhancer. *J. Ayurveda Integ. Med.* **2007**, *7*, 240–241.

63. Rastogi, S. A Review of Aconite (Vatsanabha) Usage in Ayurvedic Formulations: Traditional Views and Their Inferences. *Spatula DD* **2011**, *1*, 233–244.
64. Ren, L. Y.; Zeng, S. P. Effect of Radix *Aconiti Praeparata* Extract on Cell Apoptosis in Transplanted Liver Cancer H22. *Henan Tradit. Chin. Med.* **2008**, *28*, 34–37.
65. Sen, S. S-nitrosylation Process Acts as a Regulatory Switch for Seed Germination in Wheat. *Am. J. Plant Physiol.* **2010**, *5*, 122–132.
66. Shah, N. C. Conservation Aspects of Aconitum Species in the Himalayas with Special Reference to Uttaranchal (India). In: *Newsletter of the Medicinal Plant Specialist Group of the IUCN Species Survival Commission;* Shah, N.C. (Ed.); IUCN Species Survival Commission: Gland and Cambridge, UK, 2005; pp. 9–14.
67. Shah, R. K.; Kenjale, R. D.; Shah, D. P.; Sathaye, S.; Kaur, H. Evaluation of cardiotoxicity of Shodhit and Ashodhit samples of aconite root.*International Journal of Pharma and Biosciences*, **2010**, *7*, 65–68.
68. Sharma, S.; Rasa Tarangini. New Delhi: Motilal Banarsi Das & Co., 1982, pp. 115–116.
69. Sheokand, A.; Sharma, A.; Gothecha, V. K. Vatsanabha (*Aconitum Ferox*): From Visha to Amrita. *Int.. Ayurvedic Herb.Med.* **2012**, *2*, 423–426.
70. Shi, X. L.; Yu, X. C.; Cui, H. F.; Wu, Q.; Sun, M. J.; Huang, Y. Positive Inotropic Effect of Higenamine and 6-gingerol on Cardiac Myocytes. *Chin. J. Exp. Tradit. Med. Formulae* **2013**, *19*, 208–211.
71. Shim, S. H.; Kim, J. S.; Kang, S. S. Norditerpenoid Alkaloids from the Processed Tubers of*Aconitum carmichaelii*. *Chem. Pharm. Bull.* **2003**, *51*,999–1002.
72. Shim, S. H.; Kim, J. S.; Son, K. H.; Bae, K. H.; Kang, S. S. Alkaloids from the Roots of*Aconitum pseudo-lave*var. erectum. *J. Nat. Prod.* **2006,** *69*, 400–402.
73. Singhuber, J.; Zhu, M.; Prinz, S.; Kopp, B. *Aconitum* in Traditional Chinese Medicine: Valuable Drug or an Unpredictable Risk? *J. Ethnopharmacol.* **2009,** *126*, 18–30.
74. Song, E. F.; Sun, W. L.; Zhang, Y.; Mei, S. S. Research on clinical application and safety evaluation of Fuzi. *China J. Tradit. Chin. Med. Pharm.* **2013**, *28*, 895–901.
75. Srivastava, N.; Sharma, V.; Dobriyal, A. K.; Kamal, B.; Gupta, S.; Jadon, V. S. Influence of Pre-Sowing Treatments on *in vitro* Seed Germination of Ativisha (*Aconitum heterophyllum* Wall) of Uttarakhand—India. *Biotechnol.* **2011,** *10*, 215–219.
76. Subash, A. K.; Augustine, A. Hypolipidemic Effect of Methanol Fraction of*Aconitum heterophyllum*Wall ex Royle and the Mechanism of Action in Diet-Induced Obese Rats. *J. Adv. Pharm. Technol. Res.* **2012**, *3*, 224–228.
77. Sun, G. B.; Sun, H.; Meng, X. B.; Hu, J.; Zhang, Q.; Liu, B.; Wang, M.; Xu, H. B.; Sun, X. B. Aconitine-Induced Ca^{2+} Overload Causes Arrhythmia and Triggers Apoptosis Through p38 MAPK Signaling Pathway in Rats. *Toxicol. Appl. Pharm.***2014**, *279*, 8–22.
78. Sun, T.; Du, G. J.; Zhang, Y. P.; Li, J. H.; Liu, W. J.; Wang, Y. Y. Effect of*Aconiti Lateralis, Radix Praeparata* and *Taraxaci herba*on Chinese Medicine Signs and Symptoms of Urethane-Induced Lung Cancer in Mice. *China J. Chin. Materia Medica* **2012**, *37*, 3097–3101.
79. Trestrail III, J. H. *Criminal Poisoning: Investigational Guide for Law Enforcement, Toxicologists, Forensic Scientists and Attorneys.* Humana Press: New Jersey, USA, 2007; p. 162.
80. Turkington, C.; Mitchell, D. R. *The Encyclopedia of Poisons and Antidotes: Facts on File Library of Health and Living.*3rd Edition,Kindle Edition; Amazon Digital Services LLC, October 2009; p. 324.
81. Ukani, M. D.; Mehta, N.; Nanavati, D. *Aconitum heterophyllum* (Ativisha) in Ayurveda. *Ancient Science of Life*, **1996**, *16*, 166–171.

82. El-Shazly, M.; Tai, C. J.; Wu, T. Y.; Csupor, D.; Hohmann, J.; Chang, F. R.; Wu, Y. C. Use, History, and Liquid Chromatography/Mass Spectrometry Chemical Analysis of *Aconitum*. *J. Food Drug Anal.* **2016**, *24*, 29–45.

83. Ved, D.; Saha, D.; Ravikumar, K.; Haridasan, K. *Aconitum heterophyllum*. *The IUCN Red List of Threatened Species*, 2015, e-article: T50126560A79579556.

84. Verma, S.; Ojha, S.; Raish, M. Anti-inflammatory Activity of *Aconitum heterophyllum*on Cotton Pellet-Induced Granuloma in Rats. *J. Med. Plant Res.***2010**, *4*, 1566–1569.

85. Wang, D. P.; Lou, H. Y.; Huang, L.; Hao, X. J.; Liang, G. Y.; Yang, Z. C.; Pan, W. D. Novel Franchetine Type Norditerpenoid Isolated from the Roots of*Aconitum carmichaelii* Debx. with Potential Analgesic Activity and Less Toxicity. *Bioorg. Med. Chem. Lett.***2012**, *22*, 4444–4446.

86. Wang, L. Y. Observation at Safety of Clinical Application of Fuzi. *Chin. Mod. Med.* **2011**, *18*, 84–85.

87. Wang, L. Y.; Zhang, D. F.; Qu, X. B.; Zhang, Z. R.; Wang, Y. C. Experimental Research on Cardiotonic Effect of the Effective Parts of Crude Fuzi and Processed Fuzi. *Chin. J. Chin. Materia Medica* **2009**, *34*, 596–599.

88. Wang, T. D.; Liu, J.; Qu, L. M. Effect of Radix *Aconiti Lateralis Preparta* in Rat Models of Neuropathic Pain. *Chin. Arch. Tradit. Chin. Med.***2010**, *28*, 1083–1085.

89. Xu, Q. Y.; Yu, L. S.; Zhang, X. L.; Chen, R. M.; Chen, C. M. Effects of Fuzi and Wu Zhuyu on Coagulation System and Formation of Experimental Thrombus. *Northwest Pharm. J.***1900**, *5*, 9–11.

90. Xu, X. L.; Yang, L. J.; Jiang, J. G. Renal Toxic Ingredients and Their Toxicology from Chinese Traditional Medicine. *Expert Opin. Drug Met. Toxicol.* **2016**, *12*, 149–159.

91. Yu, L.; Wu, W. K. The Effect of Fuzi Polysaccharides on the Glucose Intake of Insulin-Resistance Adipocytes and the Possible Mechanism. *Asia Pac. Tradit. Med.* **2009**, *5*, 11–13.

92. Zhang, J.; Huang, Z. H.; Qiu, X. H.; Yang, Y. M.; Zhu, D. Y.; Xu, W. Neutral Fragment Filtering for Rapid Identification of New Diester-Diterpenoid Alkaloids in Roots of*Aconitum carmichaeli*by UHP-LC Coupled with LIT- OMS. *PLoS One* **2012**, *7*, 1–12.

93. Zhang, M.; Zhang, Y.; Chen, H. H.; Men, X. L.; Zhao, J.; Lin, H. L. The Effective Component of*Aconitum carmichaelii Debx.* for Antiventricular Arrhythmias. *Li Shizhen Med. Materia Medica Res.***2000**, *11*, 193–194.

94. Zhao, D.; Wang, J.; Cui, Y.; Wu, X. Pharmacological Effects of Chinese Herb Aconite (Fuzi) on Cardiovascular System. *J. Tradit. Chin. Med.***2012**, *32*, 308–313.

95. Zhao, J. N.; Yang, M.; Chen, Y. X. Formation and Development of Toxicity Theory of Chinese Traditional Medicine. *Chin. J. Chin. Materia Medica* **2010**, *35*, 922–927.

96. Zhao, Z.; Liang, Z.; Chan, K.; Lu, G.; Lee, E. L. M.; Chen, H.; Li, L. Unique Issue in the Standardization of Chinese Materia Medica: Processing. *Planta Medica* **2010**, *76*, 1975–1986.

97. Zhou, G.; Tang, L.; Zhou, X.; Wang, T.; Kou, Z.; Wang, Z. A. Review on Phytochemistry and Pharmacological Activities of the Processed Lateral Root of *Aconitum carmichaelii* Debeaux. *J. Ethnopharmacol.* **2015**, *160*, 173–193.

98. Zhou, Y. P.; Liu, W. H. Review and Reappraise on Researches of Fuzi's Effect on Cardio-vascular System—Part I. *Pharmacol. Clin. Chin. Materia Medica***2013**, *29*, 198–205.

PART III
Role of Medicinal Plants in Disease Management

CHAPTER 7

ROLE OF NATURAL POLYPHENOLS IN OXIDATIVE STRESS: PREVENTION OF DIABETES

BRAHM KUMAR TIWARI and KANTI BHOOSHAN PANDEY

ABSTRACT

The impact of excess calorie intake and reduced physical activity has been seen into exponential rise in obesity and prevalence of diabetes all over the world. The chronic hyperglycemia due to prolonged increase in blood glucose level due to defects in either insulin action/secretion or both results into dysfunction/long-term damage/failure of various vital organs, especially eyes, kidney, heart, and brain. Oxidative stress plays a crucial role in the pathogenesis of diabetes by mediating metabolic activities such as glucose oxidation, non-enzymatic oxidation of proteins, oxidative degradation of glycated proteins, and simultaneously challenges inherent defense mechanisms. Plant polyphenols are secondary metabolites that occur largely in fruits and vegetables, and have been reported to elicit medicinal properties. Various animal and clinical studies suggest that these bioactive compounds play an important role in the prevention of many chronic human diseases and promoting overall human health. In the present chapter, role and emerging mechanisms of plant polyphenols in intervening oxidative stress mediated diabetic complications have been described.

7.1 INTRODUCTION

Diabetes mellitus (DM) is a *heterogeneous* group of metabolic disorder characterized by elevated glucose level in blood due to defects in either insulin action or insulin secretion and sometimes both. It has been reported that the chronic hyperglycemia negatively influences vital organs of the body including eyes, kidneys, blood vessels, nerves, and heart and makes them malfunctioned.[1,2,63] People with DM often develop an array of vascular and neuropathic complications that seriously erode the quality and span of life.[11] In the current scenario, the prevalence of diabetes, mainly Type 2 diabetes (T2D), has been evolved as a global public health threat. It has been proposed that the prevalence of DM among adults of 20-70 years of age range will rise from 285 million in 2010 to 438 million by the year 2030, which is approximately double.[87] Type 1 diabetes (T1D) and T2D are characterized by progressive failure of b-cell function, in which T1D is caused by an autoimmune assault against the b cells, inducing its progressive damage and finally death.[54] However, the development and progression of T2D are variable, consisting of asymptotes degrees of b-cell failure relative to varying degrees of insulin resistance.[22]

A wide dimension of therapies available for the treatment of diabetes including oral anti-diabetic agents, such as biguanides, glinides, metformin, sulfonylureas, and insulin; however, most of them have serious adverse side effects.[38,69] Conventional treatments including Ayurveda against diabetes and associated pathologies advocate use of plant and plant-derived drugs as potent alternative medicine because of their easy availability, low cost, and relatively low side effects. The major biological activities of herbal drugs derived from medicinal plants are their pancreatic β-cell regenerating and insulin-releasing properties.[77]

Polyphenols are naturally occurring secondary metabolites found abundantly in plant-derived drinks, such as coffee, tea, wine, vegetables, legumes, fruits, and cereals.[64, 92] Several experiments over the last two decades have confirmed that plant polyphenols pass strong antioxidant potential that may responsible for health-promoting effects of natural products/herbal remedies.[31,47] An array of report documents that polyphenols effectively work in the prevention of many life-threatening diseases, such as vascular disorders, various types of cancers, neurodegenerative impairments, and aging.[32,65,79]

In this chapter, the role and emerging mechanisms of plant-derived polyphenols in intervening diabetic complications have been described.

7.2   OXIDATIVE STRESS IN THE DEVELOPMENT OF DM

Oxidative stress may be addressed as a measure of balance between production of reactive oxygen species (ROS) and their neutralization in living environment. Free radicals or ROS generate during metabolic activities in the body and are treated as a necessary evil in a biological system.[5] They are highly reactive and play dual role: both toxic and beneficial. ROS inside the body are involved in different vital functions, such as cell growth regulation, phagocytosis, energy production, and intercellular signaling. Under normal conditions, various antioxidant systems involved in restoring the redox balance and maintained the low level of intracellular ROS. However, since humans are constantly exposed to different types of radiations, pollutants, smoke, endogenous respiratory burst and enzymatic reactions, the generation of ROS in the body get enhanced and cellular redox balance get impaired.[30,42] In this context, oxidative stress can also be characterized as imbalance between the pro-oxidants and antioxidants in a cellular system.[9,73] According to finding of many researchers, oxidative stress is the major culprit behind progression and development of many deleterious pathological events including cancer, arthritis, atherosclerosis, neuro-problems, and hyperglycemia followed by DM.[28,67]

Influence of oxidative stress in pathogenesis of both T1D and T2D is very vital. ROS are formed in a disproportionate manner during diabetes as a result of different metabolic activities, such as impaired oxidation of glucose, glycation of proteins by non-enzymatic means, and subsequently their oxidative degradation.[53,57] Excessive generation of ROS may cause altered cellular physiology by deterioration of cellular organelles, involved in redox homeostasis via peroxidation of lipids, malfunctioning enzymes, and promoting the insulin resistance. All these ramifications of severe oxidative stress can induce the progression of diabetic pathologies.[54,80]

A number of mechanisms involved in hyperglycemia-mediated tissue damage have been reported with experimental evidence.[16,93] For inducing damage to the tissues by hyperglycemia, the major research studies have emphasized that there are five main events: (1) enhanced flux of $C_6H_{12}O_6$ and related sugars through polyol pathway; (2) elevated generation of advanced glycation end products (AGEs); (3) overexpression of the receptors for AGEs and its activating ligands; (4) protein kinase C (PKC) isoforms activation; and (5) overactivation of hexosamine pathway.[10,29,52] Many research studies indicate that excess generation of ROS by mitochondria as a single upstream event that activates all of these five events of hyperglycemia induced complications.[16,93]

7.3 POLYPHENOLS AS NOVEL THERAPEUTIC AGENTS AGAINST DM

Polyphenols are widely occurring secondary plant metabolites. Thousands of the investigated polyphenols have reported to be active to promote human health.[64] Structurally, polyphenols may be simple like phenolic acids or complex polymerized like tannins. Flavonoids represent the ubiquitously occurring largest class of polyphenols.[14] Major polyphenols that commonly occur in the diet are: anthocyanins, flavanols (including catechins, proanthocyanidins, tannins, thearubigins), phenolic acids, ellagitannins, isoflavones, flavanones, and flavones.[20,21,71,78,85]

Polyphenols in particular have been evaluated throughout the world for their diverse medicinal properties. Clinical trials on animals and humans have revealed that polyphenols as natural phytocompounds may contribute in protection and promotion of overall human health.[68,83,84] Increasing evidence suggests that chronic oxidative mediated age associated complications can be prevented by the regular consumption of polyphenols or polyphenol rich diet.[56,89,90]

Despite being potent antioxidant, plant polyphenols possess many other dynamic biological properties, such as anti-allergic, analgesic, anti-microbial, anti-mutagenic, anti-viral, anti-inflammatory, anti-atherogenic, anti-thrombotic, and anti-hyperglycemia.[26,43,94] There are evidences that plant-derived polyphenols can modulate the process of immunity, apoptosis, and cell cycle regulation. They have also been reported to induce antioxidative enzymes and in some cases modulation of nuclear factor-kappa B (NF-κB) signaling and other vital pathways.[35,66,81]

7.4 MECHANISMS OF ACTION: PROTECTIVE EFFECTS OF POLYPHENOLS AGAINST DIABETIC COMPLICATIONS

Various studies investigating anti-diabetic effects of polyphenols report that polyphenols may provide protection against DM via multiple pathways including modulating digestion of carbohydrate, secretion, and sensitivity of insulin.[6] Some important pathways modulated by polyphenols are being discussed in this section.

7.4.1 MODULATION OF INTESTINAL GLUCOSE UPTAKE

Compromised carbohydrate metabolism with increasing insulin resistance is the main metabolic impairment in DM leading to hyperglycemia, which results into alteration in their digestion followed by absorption, exhaustion in storage of glycogen, elevated gluconeogenesis and insulin resistance of peripheral tissue, β-cell dysfunction, over output hepatic glucose, and defect in insulin signaling pathways.[25,37]

The potential efficiency and efficacy of many polyphenols on the metabolism of carbohydrate followed by glucose homeostasis have been evaluated in animal models and many clinical trials.[36] Supplementation of appropriate amount of polyphenols is reported to balance blood glucose level in diabetics.[40,75] It has been postulated that a convenient dietary mechanism may be provided by polyphenols that act behind polyphenol-mediated regulation of absorption of sugar in intestine. In the long-term management of diabetes and associated complications, this mechanism may be an effective factor.[44]

Among many mechanisms proposed to describe the regulation of blood glucose level by polyphenols, reduced intestinal glucose absorption caused by polyphenols is most accepted once.[44,75] The physiological studies have provided evidence that sodium-dependent glucose co-transporter, expressed in the jejunum, involves in the transport of glucose in epithelial cells and from there the glucose is carried in blood via facilitated glucose transporter (GULT). There is experimental evidence that interruption of glucose uptake via sodium-dependent glucose co-transporter in the small intestine may prevent the prolonged condition of hyperglycemia that later develops as diabetes.[40]

Excessive post-prandial glucose excursion is controlled through the inhibition of activities of digestive enzymes and membrane transporters involved in glucose metabolism.[86] Plant polyphenols are also reported to modulate intestinal a-glucosidase and a-amylase activities that are crucial in the development of hyperglycemia.[59] Different polyphenols, such as 5-caffeoylqunic acid, gallate esters, theaflavins, and proanthocyanidins, have been noted to inhibit a-amylase activity. Some polyphenols including green tea catechins inhibit lactase thereby promotes delayed generation of glucose. Flavonols and phlorizin are reported to inhibit SGLT1 and GLUT2 transporters, thus prevent the absorption of glucose.[86,91]

7.4.2   IMPROVEMENT IN INSULIN SENSITIVITY

Disorders such as obesity and diabetes with cardiovascular impairments are resultant of metabolic fate such as insulin resistance or insulin insensitivity and endothelial dysfunction. In this context, therapies that target either insulin resistance or endothelial dysfunction are supposed to protect metabolic as well as cardiovascular disorders simultaneously and thus disease outcomes during diabetes.[3,60]

Several studies on intervention of diabetes and its complications have reported that polyphenols may improve insulin sensitivity through the factors like increased insulin-dependent glucose uptake via GLUT-4, translocation of cell membrane, activation of adenosine 5'-monophosphate-activated protein kinase (AMPK), and phosphoinositide 3-kinase *inhibitor* (PI3K).[4,51]

Studies provide results that many plant polyphenols may modulate insulin sensitivity such as green tea catechins like Epigallocatechin-3-gallate (EGCG)[15,50], chlorogenic acid, 4-caffeoylquinic, gallic acid, 5-hydroxy-methylfurfural, protocatechuic acid[70], cinnamon polyphenols (catechin, epicatechin)[72], anthocyanin (cyanidin 3-glucoside)[76], resveratrol[13], olive leaf polyphenols (vpigenenin, flavonoid, verbascoside, oleic acid, quercetin, luteolin, rutin, oleuropein, hydroxytyrosol, and kaempferol).[24]

In skeletal muscles, glucose uptake is promoted by insulin by activating PI3K thereby enhanced translocation of GLUT4 to the plasma membrane.[61] AMPK is reported as a major metabolic-sensing protein involved in the protection of metabolism-related disorders. AMPK can be activated by many polyphenols including berberine, epigallocatechin gallate, resveratrol, and quercetin. Activation of AMPK can induce the generation of ATP (adenosine triphosphate) through pathways including glycolysis and b-oxidation. Irrespective to these ATP-consuming pathways, synthesis of fatty acid as well as cholesterol and gluconeogenesis are reported to be suppressed by AMPK activation.[33,41,95] Studies have revealed the potential effect of polyphenols in the activation of AMPK during diabetes.[8,97] Resveratrol and chlorogenic acid have been reported to increase the phosphorylation of AMPK, thereby contributing to improvements in whole-body glucose and lipid metabolism.[27]

7.4.3   ENHANCEMENT IN THE B-CELL FUNCTION

In T1D, the degeneration of β cells of pancreas mediated by auto-immune response causes absolute insulin scarcity and development of hyperglycemia. In T2D, both b-cell dysfunction as well as insulin resistance contribute in the development of hyperglycemia.[18] Thus, enhancing β-cell function to retard the progression of both types of diabetes, type 1 and 2, and their related complications may be one of the potent strategies against the development of these types of diabetes.

The therapeutic effect of polyphenols on β-cell function has been reported with focus on the mechanism of action in many studies.[17,23] It has been highlighted by many researchers that the targeted action of many plant polyphenols on pancreatic β cells stimulates the secretion of insulin via activating specific cellular targets and thus protects β cells from inflammation and oxidative stress induced degeneration (Figs. 7.1–7.3).[39,74]

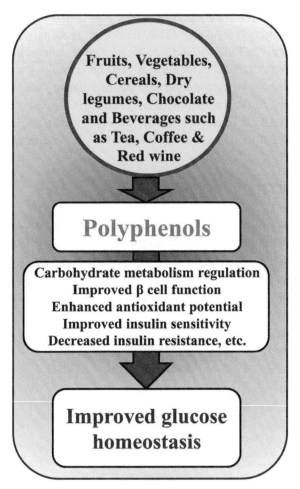

FIGURE 7.1   Role of plant-derived polyphenols to manage blood glucose level during diabetes.

Polyphenols may improve the β-cell function by[7,55,58,82,96]: (1) protecting β cells against oxidative damage, (2) delaying β-cell failure, (3) inhibiting β-cell apoptosis, and (4) modulating insulin production and secretion. There are experimental evidences that some polyphenols (resveratrol, genistein, and quercetin) inhibit phosphodiesterase (PDE) activity and enhance β-cell function that leads to glucose tolerance in humans and animal models, in turn, providing protection against diabetes.[74] In other studies, polyphenolic compounds have been reported to directly modulate the function of b cells.[62] Investigating the molecular targets of plant polyphenols especially in diabetes is challenging. Similar to currently available anti-diabetic drugs, supplementation with polyphenols modulates activities of many signaling proteins, such as including protein kinases, ion channels, and/or transcription factors.[88] Based on the experimental evidences, it may be concluded that plant polyphenols can improve b-cell function not only by inducing the survival factors or effectors of insulin secretion, but also by reducing the activities of undesirable cellular events[49]: lipid peroxidation, protein oxidation, cardiovascular dysfunction, and other damaging effect of metabolic syndrome.

7.4.4   REGULATION OF GLUCOSE METABOLISM

In the continuation of previously discussed glucose uptake modulating the effect of polyphenols, the active biocompounds have also been reported to regulate glucose metabolism by decreasing gluconeogenesis and increasing glycogenesis in liver, increasing glycolysis and glucose oxidation, increased hepatic glucokinase activity, and glycogen content.[46]

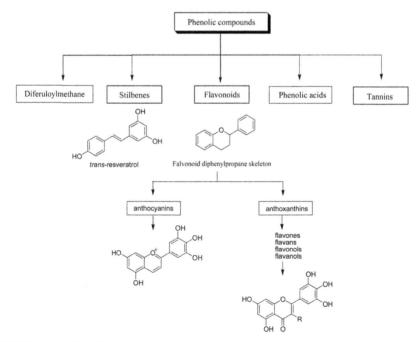

FIGURE 7.2   Classification of dietary polyphenols.

Source: Ref. [35]; https://doi.org/10.3390/i8090950, Open Source Article.

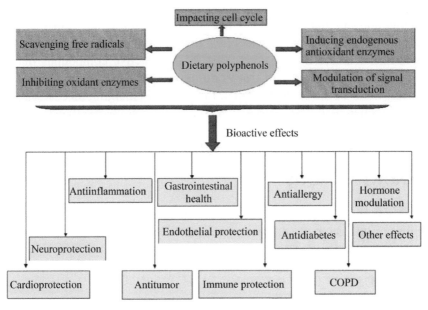

FIGURE 7.3   Bioactivities of dietary polyphenols.

Source: Ref. [35]; https://doi.org/10.3390/i8090950, Open Source Article.

Polyphenols may normalize the activities of hepatic gluconeogenic enzymes, hexokinase, glucose-6-phosphatase, fructose-1,6-bisphosphatase, and phosphofructokinase, which may be elevated in diabetes and as a result, there is a rise in the glucose production during diabetes in the liver.[48] Diets rich in polyphenols such as green tea, red wine, grains, carrots, berries, fresh fruit jam, olive oil, and vegetables, including onion, soybeans, spinach, and buckwheat may improve glucose metabolism in the people at high risk for the development of diabetes and related disorders.[12,19,35,45]

7.5   SUMMARY

The impact of excess calorie intake and reduced physical activity has been observed into exponential rise in obesity and prevalence of diabetes all over the world. The chronic hyperglycemia due to prolonged increase in blood glucose level due to defect in either insulin action/secretion or both may result into dysfunction/damage of various vital organs such as eyes, kidney, vessels, and brain. Oxidative stress plays a crucial role in the pathogenesis of diabetes by mediating metabolic activities such as: oxidation of glucose, non-enzymatic degeneration of proteins, oxidative degradation of glycated proteins, and simultaneously challenges inherent defense mechanisms. Plant polyphenols are secondary metabolites that are found largely in fruits and vegetables. Clinical trials suggest that these bioactive compounds play a vital role in the prevention of many chronic diseases and promoting overall human health. In the present chapter, the role and emerging mechanisms of plant polyphenols in the intervention of oxidative stress mediated diabetic complications have been described.

ACKNOWLEDGMENTS

K.P.B. gratefully acknowledges the Director of CSIR-CSMCRI, Bhavnagar for his constant support and encouragement.

KEYWORDS

- diabetes
- oxidative stress
- polyphenols
- phytomedicine

REFERENCES

1. American Diabetes Association (ADA). Diagnosis and Classification of Diabetes Mellitus. *Diabetes Care* **2012,** *35* (1), 64–71.
2. American Diabetes Association (ADA). Diagnosis and Classification of Diabetes Mellitus. *Diabetes Care* **2010,** *33* (1), 62–69.
3. Anderson, R. A. Chromium and Polyphenols from Cinnamon Improve Insulin Sensitivity. *Proceed. Nutr. Soc.* **2008,** *67* (1), 48–53.
4. Anderson, R. A.; Broadhurst, C. L.; Polansky, M. M.; Schmidt, W. F.; Khan, A.; Flanagan, V. P.; Schoene, N. W.; Graves, D. J. Isolation and Characterization of Polyphenol Type-A Polymers from Cinnamon with Insulin-Like Biological Activity. *J. Agric. Food Chem.* **2004,** *52* (1), 65–70.
5. Anhe, F. F.; Desjardins, Y.; Pilon, G.; Dudonne, S.; Genovese, M. I.; Lajolo, F. M.; Marette, A. Polyphenols and Type-2 Diabetes: A Prospective Review. *Pharm. Nutr.* **2013,** *1* (4), 105–114.

6. Babu, P. V.; Liu, D.; Gilbert, E. R. Recent Advances in Understanding the Anti-Diabetic Actions of Dietary Flavonoids. *J. Nutr. Biochem.* **2013**, *24,* 1777–1789.

7. Bae, U. J.; Jang, H. Y.; Lim, J. M.; Hua, L.; Ryu, J. H.; Park, B. H. Polyphenols Isolated from *Broussonetia kazinoki* Prevent Cytokine-Induced β-Cell Damage and the Development of Type-1 Diabetes. *Exp. Mol. Med.* **2015,** *47* (3), 1–10.

8. Bahadoran, Z.; Mirmiran, P.; Azizi, F. Dietary Polyphenols as Potential Nutraceuticals in Management of Diabetes: A Review. *J. Diabetes Metab. Dis.* **2013**, *12* (1), 43.

9. Baynes, J. W. Role of Oxidative Stress in Development of Complications in Diabetes. *Diabetes,* **1991**, *40* (4), 405–412.

10. Bloomer, R. J.; Goldfarb, A. H.; Mckenzie, M. J. Oxidative Stress Response to Aerobic Exercise: Comparison of Antioxidant Supplements. *Med. Sci. Sports Exer.* **2006**, *38,* 1098–1105.

11. Boyle, J. P.; Thompson, T. J.; Gregg, E. W.; Barker, L. E.; Williamson, D. F. Projection for 2050: Burden of Diabetes in the US Adult Population - Dynamic Modeling of Incidence, Mortality, and Prediabetes Prevalence. *Popul. Health Metrics* **2010**, *8*, 29.

12. Bozzetto, L. *Diets Naturally-Rich in Polyphenols Improve Glucose Metabolism in People at High Cardiovascular Risk: A Controlled Randomized Trial*; *EAS*: Madrid, Spain; June 1, 2014; Abstract M087.

13. Brasnyo, P.; Molnar, G. A.; Mohas, M.; Marko, L.; Laczy, B.; Cseh, J.; Mikolas, E. Resveratrol Improves Insulin Sensitivity, Reduces Oxidative Stress and Activates the Akt Pathway in Type-2 Diabetic Patients. *Br. J. Nutr.* **2011**, *106* (3), 383–389.

14. Bravo, L. Polyphenols: Chemistry, Dietary Sources, Metabolism, and Nutritional Significance. *Nutr. Rev.* **1998**, *56*, 317–333.

15. Brown, A. L.; Lane, J.; Coverly, J.; Stocks, J.; Jackson, S.; Stephen, A.; Bluck, L.; Coward, A.; Effects of Dietary Supplementation with the Green Tea Polyphenol Epigallocatechin Gallate on Insulin Resistance and Associated Metabolic Risk Factors: Randomized Controlled Trial. *Br. J. Nutr.* **2009**, *101* (6), 886–894.

16. Brownlee, M. The Pathobiology of Diabetic Complications: Unifying Mechanism. *Diabetes* **2005**, *54* (6), 1615–1625.

17. Buscemi, S.; Marventano, S.; Antoci, M.; Cagnetti, A.; Castorina, G.; Coffee and Metabolic Impairment: An Updated Review of Epidemiological Studies. *NFS J.* **2016**, *3*, 1–7.

18. Cerf, M. E. Beta Cell Dysfunction and Insulin Resistance. *Front. Endocrinol. (Lausanne)* **2013**, *4*, 1–12.

19. Chiva-Blanch, G.; Urpi-Sarda, M.; Ros, E.; Valderas-Martinez, P.; Casas, R.; Arranz, S.; Guillen, M. Effects of Red Wine Polyphenols and Alcohol on Glucose Metabolism and Lipid Profile: A Randomized Clinical Trial. *Clin. Nutr.* **2013**, *32* (2), 200–206.

20. Clifford, M. N. Anthocyanins: Nature, Occurrence and Dietary Burden. *J. Sci. Food Agric.* **2000**, *80*, 1063–1072.

21. Clifford, M. N. Chlorogenic Acids and Other Cinnamates: Nature, Occurrence and Dietary Burden. *J. Sci. Food Agric.* **1999**, *79*, 362–372.

22. Cnop, M.; Welsh, N.; Jonas, J. C.; Jorns, A.; Lenzen, S.; Eizirik, D. L. Mechanisms of Pancreatic Beta-Cell Death in Type-1 and Type-2 Diabetes: Many Differences, Few Similarities. *Diabetes* **2005**, *54* (2), 97–107.

23. Dall'Asta, M.; Bayle, M. Protection of Pancreatic β-Cell Function by Dietary Polyphenols. *Phytochem. Rev.* **2015**, *14*, 933–959.

24. De Bock, M.; Derraik, J. G.; Brennan, C. M.; Biggs, J. B.; Morgan, P. E.; Hodgkinson, S. C. Olive (*Olea europaea* L.) Leaf polyphenols Improve Insulin Sensitivity In

Middle-Aged Overweight Men: A Randomized, Placebo-Controlled, Crossover Trial. *PLoS One* **2013,** *8* (3), 1–8.

25. Dinneen, S.; Gerich, J.; Rizza, R. Carbohydrate Metabolism In Non-Insulin-Dependent *Diabetes Mellitus*. *New Engl. J. Med.* **1992,** *327,* 707–713.

26. Edziri, H.; Mastouri, M.; Aouni, M.; Verschaeve, L. Polyphenols Content, Antioxidant And Antiviral Activities Of Leaf Extracts Of *Marrubium Deserti* Growing In Tunisia. *South African J. Bot.* **2012,** *80,* 104–109.

27. Egawa, T.; Tsuda, S.; Oshima, R.; Goto, K.; Hayashi, T. Activation Of 5′AMP-Activated Protein Kinase In Skeletal Muscle By Exercise And Phytochemicals. *J. Sports Med. Phys. Fit.* **2014,** *3* (1), 55–64.

28. Feher, J.; Csomos, G.; Verekei, A. *Free Radical Reactions in Medicine*; 1st ed.; Springer Verlag: Germany, 1987, p 199.

29. Giacco, F.; Brownlee, M. Oxidative Stress and Diabetic Complications. *Circ. Res.* **2010,** *107* (9), 1058–1070.

30. Gilgun-Sherki, Y.; Melamed, E.; Offen, D. Oxidative Stress Induced- Neurodegenerative Diseases: Need For Antioxidants That Penetrate The Blood Brain Barrier. *Neuropharmacology* **2001,** *40* (8), 959–975.

31. Greiner, A. K.; Papineni, R. V. L.; Umar, S. Chemoprevention in Gastrointestinal Physiology And Disease: Natural Products And Microbiome. *Am. J. Physiol. Gastrointest. Liver Physiol.* **2014,** *307* (1), 1–15.

32. Gutowski, M.; Kowalczyk, S. A. Study Of Free Radical Chemistry: Their Role And Pathophysiological Significance. *Acta Biochem. Polon.* **2013,** *60* (1), 1–16.

33. Hajiaghaalipour, F.; Khalilpourfarshbafi, M.; Arya, A. Modulation of Glucose Transporter Protein By Dietary Flavonoids In Type-2 *Diabetes Mellitus. Int. J. Biol. Sci.* **2015,** *11* (5), 508–524.

34. Hollman, P. C.; van Trijp, J. M. Relative Bioavailability Of The Antioxidant Flavonoid Quercetin From Various Foods In Man. *FEBS Lett.* **1997,** *418,* 152–156.

35. Han, X.; Shen, T.; Lou, H. Dietary Polyphenols and Their Biological Significance. *Int. J. Mol. Sci.* **2007,** *8* (9), 950–988. Open source.

36. Hanhineva, K.; Torronen, R.; Bondia-Pons, I.; Pekkinen, J.; Kolehmainen, M. Impact of Dietary Polyphenols on Carbohydrate Metabolism. *Int. J. Mol. Sci.* **2010,** *11,* 1365–1402.

37. Harcourt, B. E.; Penfold, S. A.; Forbes, J. M. Coming Full Circle In *Diabetes Mellitus*: From Complications To Initiation. *Nat. Rev. Endocrinol.* **2013,** *9* (2), 113–123.

38. Hong, S. H.; Heo, J. I.; Kim, J. H.; Kwon, S. O.; Yeo, K. M.; Bakowska-Barczak, A. M.; Kolodziejczyk, P. Antidiabetic and Beta Cell Protection Activities Of Purple Corn Anthocyanins. *Biomol. Ther.* **2013,** *21* (4), 284–289.

39. Hong, S.H.; Woo, M.; Kim, M.; Song, Y.O. Hypolipidemic and Antidiabetic Effects Of Functional Rice Cookies In High-Fat Diet-Fed ICR Mice and db/db Mice. *J. Med. Food* **2018,** *21* (6), 535–543.

40. Hossain, S. J.; Kato, H.; Aoshima, H.; Yokoyama, T.; Yamada, M.; Hara, Y. Polyphenol-Induced Inhibition Of The Response Of Na(+)/ Glucose Cotransporter Expressed In Xenopus Oocytes. *J. Agric. Food Chem.* **2002,** *50* (18), 5215–5219.

41. Hwang, J. T.; Kwon, D. Y. AMP-Activated Protein Kinase: Potential Target For The Diseases Prevention By Natural Occurring Polyphenols. *New Biotechnol.* **2009,** *26* (1–2), 17–22.

42. Jaganjac, M.; Tirosh, O.; Cohen, G.; Sasson, S.; Zarkovic, N. Reactive Aldehydes-Second Messengers Of Free Radicals In Diabetes Mellitus. *Free Radic. Res.* **2013,** *47* (1), 39–48.

43. Jangga, R. N.; Assad, S.; Bukhari, A. Effect of Polyphenols Klika Ongkea (*Mezzetia Parviflora Becc*) Against Blood Glucose Wistar Rats Induced By Streptozotocin. *Int. J. Sci. Technol. Res.* **2015,** *4* (4), 307–312.

44. Johnston, K.; Sharp, P.; Clifford, M.; Morgan, L. Dietary Polyphenols Decrease Glucose Uptake By Human Intestinal Caco-2 Cells. *FEBS Lett.* **2005,** *579* (7), 1653–1657.

45. Josic, J.; Olsson, A. T.; Wickeberg, J.; Lindstedt, S.; Hlebowicz, J. Does Green Tea Affect Postprandial Glucose, Insulin And Satiety In Healthy Subjects: A Randomized Controlled Trial? *Nutr. J.* **2010,** *9*, 1–8.

46. Jung, U. J.; Lee, M. K.; Park, Y. B.; Jeon, S. M.; Choi, M. S. Antihyperglycemic And Antioxidant Properties Of Caffeic Acid in db/db Mice. *J. Pharmacol. Experiment. Ther.* **2006,** *318* (2), 476–483.

47. Kasote, D. M.; Katyare, S. S.; Hegde, M. V.; Bae, H. Significance of Antioxidant Potential Of Plants And Its Relevance To Therapeutic Applications. *Int. J. Biol. Sci.* **2015,** *11* (8), 982–991.

48. Kazeem, M. I.; Akanji, M. A.; Yakubu, M. T.; Ashafa, A. O. Protective Effect Of Free And Bound Polyphenol Extracts From Ginger (*Zingiber Officinale Roscoe*) On The Hepatic Antioxidant And Some Carbohydrate Metabolizing Enzymes Of Streptozotocin-Induced Diabetic Rats. *Evidence Based Complement. Alternat. Med.* **2013,** *2013*, Article ID: 935486, 7.

49. Kerimi, A.; Williamson, G. At The Interface Of Antioxidant Signaling And Cellular Function: Key Polyphenol Effects. *Mol. Nutr. Food Res.* **2016,** *60*, 1770–1788.

50. Kim, J. A. Mechanism Underlying Beneficial Health Effects Of Tea Catechins To Improve Insulin Resistance And Endothelial Dysfunction. *Endocrine Metab. Immune Disorders Drug Targets* **2008,** *8* (2), 82–88.

51. Kim, Y.; Keogh, J. B. Polyphenols and Glycemic Control. *Nutrients* **2016,** *8* (1), 1–27.

52. Kowluru, R. A.; Chan, P. Oxidative Stress And Diabetic Retinopathy. *Experiment. Diab. Res.* **2007,** *2007*, Article ID: 43603, 12.

53. Kumarappan, C. T.; Thilagam, E.; Vijayakumar, M.; Mandal, S. C. Modulatory Effect Of Polyphenolic Extracts Of *Ichnocarpus Frutescens* On Oxidative Stress In Rats With Experimentally Induced Diabetes. *Indian J. Med. Res.* **2012,** *136* (5), 815–821.

54. Lencioni, C.; Lupi, R.; Del Prato, S. Beta-Cell Failure In Type-2 *Diabetes Mellitus*. *Curr. Diabetes Rep.* **2008,** *8* (3), 179–184.

55. Li, C.; Allen, A.; Kwagh, J.; Doliba, N. M.; Qin, W. Green Tea Polyphenols Modulate Insulin Secretion By Inhibiting Glutamate Dehydrogenase. *J. Biol. Chem.* **2006,** *281*(15), 10214–10221.

56. Liu, R. H. Potential Synergy Of Phytochemicals In Cancer Prevention: Mechanism Of Action. *J. Nutr.* **2004,** *134*, 3479–3485.

57. Maritin, A. C.; Sanders, R. A.; Watkins, J. B. Diabetes, Oxidative Stress, And Antioxidants: A Review. *J. Biochem. Mol. Toxicol.* **2003,** *17* (1), 24–38.

58. Martín, M. A.; Fernandez-Millan, E.; Ramos, S.; Bravo, L.; Goya, L. Cocoa Flavonoid Epicatechin Protects Pancreatic Beta Cell Viability And Function Against Oxidative Stress. *Mol. Nutr. Food Res.* **2014,** *58* (3), 447–456.

59. McDougall, G. J.; Shapiro, F.; Dobson, P.; Smith, P.; Blake, A.; Stewart, D. Different Polyphenolic Components Of Soft Fruits Inhibit Alpha-Amylase And Alpha-Glucosidase. *J. Agric. Food Chem.* **2005,** *53* (7), 2760–2766.

60. Munir, K. M.; Chandrasekaran, S.; Gao, F.; Quon, M. J. Mechanisms for Food Poly-phenols To Ameliorate Insulin Resistance And Endothelial Dysfunction: Therapeutic Implications For Diabetes And Its Cardiovascular Complications. *Am. J. Physiol. Endocrinol. Metab.* **2013,** *305* (6), 679–686.

61. O'Neill, H. M. AMPK and Exercise: Glucose Uptake and Insulin Sensitivity. *Diabetes Metab. J.* **2013,** *37* (1), 1–21.

62. Oh, Y. S. Plant-Derived Compounds Targeting Pancreatic Beta Cells For The Treatment Of Diabetes. *Evidence Based Complement. Altern. Med.* **2015,** *2015,* Article ID: 629863, 12.

63. Palsamy, P.; Subramanian, S. Resveratrol, A Natural Phytoalexin, Normalizes Hypergly-cemia In Streptozotocin- Nicotinamide Induced Experimental Diabetic Rats. *Biomed. Pharmacother.* **2008,** *62* (9), 598–605.

64. Pandey, K. B.; Rizvi, S. I. Plant Polyphenols As Dietary Antioxidants In Human Health And Disease. *Oxid. Med. Cell. Longev.* **2009,** *2,* 270–278.

65. Pandey, K. B.; Rizvi, S. I. Recent Advances In Health Promoting Effect Of Dietary Polyphenols. *Curr. Nutr. Food Sci.* **2012,** *8* (4), 254–264.

66. Pandey, K. B.; Mishra, N.; Rizvi, S. I. Myricetin May Provide Protection Against Oxidative Stress In Type-2 Diabetic Erythrocytes. *Zeitschrift fur Naturforschung C* **2009,** *64* (9–10), 626–630.

67. Pandey, K. B.; Rizvi, S. I. Biomarkers Of Oxidative Stress In Red Blood Cells. *Biomed. Papers Med. Facult. Univ. Palacký, Olomouc, Czechoslovakia Republic* **2011,** *155* (2), 131–136.

68. Pandey, K.B.; Rizvi, S.I. Plant Polyphenols in Healthcare and Aging. In *Nutritional Antioxidant Therapies: Treatments and Perspectives*; Al-Gubory, K. and Laher, I., Eds.; Springer International Publishing: Cham, France, 2017; pp 267–282.

69. Patel, D. K.; Prasad, S. K.; Kumar, R.; Hemalatha, S. An Overview On Antidiabetic Medicinal Plants Having Insulin Mimetic Property. *Asian Pacific J. Tropic. Biomed.* **2012,** *2,* 320–330.

70. Peng, C. H.; Chyau, C. C.; Chan, K. C.; Chan, T. H.; Wang, C. J.; Huang, C. N. Hibiscus Sabdariffa Polyphenolic Extract Inhibits Hyperglycemia, Hyperlipidemia, And Glycation - Oxidative Stress While Improving Insulin Resistance. *J. Agric. Food Chem.* **2011,** *59* (18), 9901–9909.

71. Perez-Jimenez, J.; Neveu, V.; Vos, F.; Scalbert, A. Identification Of The 100 Richest Dietary Sources Of Polyphenols: An Application Of The Phenol-Explorer Database. *Eur. J. Clin. Nutr.* **2010,** *64,* 112–120.

72. Qin, B.; Panickar, K. S.; Anderson, R. A. Cinnamon: Potential Role In The Prevention Of Insulin Resistance, Metabolic Syndrome, And Type-2 Diabetes. *J. Diabetes Sci. Technol.* **2010,** *4* (3), 685–693.

73. Rahal, A.; Kumar, A.; Singh, V. Oxidative Stress, Pro-Oxidants, And Antioxidants: The Interplay. *BioMed Res. Int.* **2014,** *2014,* Article ID: 761264, 19.

74. Rouse, M.; Younes, A.; Egan, J. M. Resveratrol And Curcumin Enhance Pancreatic B-Cell Function By Inhibiting Phosphodiesterase Activity. *J. Endocrinol.* **2014,** *223* (2), 107–117.

75. Rzepecka-Stojko, A.; Stojko, J.; Kurek-Gorecka, A.; Gorecki, M.; Kabała-Dzik, A.; Kubina, R.; Mozdzierz, A.; Buszman, E. Polyphenols From Bee Pollen: Struc-ture, Absorption, Metabolism And Biological Activity. *Molecules* **2015,** *20* (12), 21732–21749.

76. Sasaki, R.; Nishimura, N.; Hoshino, H.; Isa, Y.; Kadowaki, M.; Ichi, T. Cyanidin 3-Glucoside Ameliorates Hyperglycemia And Insulin Sensitivity Due To Downregulation Of Retinol Binding Protein 4 Expression In Diabetic Mice. *Biochem. Pharmacol.* **2007,** *74* (11), 1619–1627.

77. Saxena, A. M.; Mukherjee, S. K.; Shukla, G., Eds. *Progress of Diabetes Research in India During 20th Century*; National Institute of Science Communication; New Delhi, 2006; p 104.

78. Scalbert, A.; Williamson, G. Dietary Intake And Bioavailability Of Polyphenols. *J. Nutr.* **2000,** *130*, 2073–2085.

79. Scalbert, A.; Manach, C.; Morand, C.; Remesy, C.; Jimenez, L. Dietary Polyphenols And The Prevention Of Diseases. *Critic. Rev. Food Sci. Nutr.* **2005,** *45* (4), 287306.

80. Sheweita, S. A.; Mashaly, S.; Newairy, A. A.; Abdou, H. M.; Eweda, S. M. Changes In Oxidative Stress And Antioxidant Enzyme Activities In Streptozotocin-Induced *Diabetes Mellitus* In Rats: Role Of *Alhagi Maurorum* Extracts. *Oxid. Med. Cell. Longev.* **2016,** *2016*, Article ID: 5264064, 8.

81. Shizue, K.; Noriko, Y.; Hideyo, S.; Kiharu, I. Anti-Diabetic Effects of *Actinidia arguta* polyphenols on Rats and KK-Ay Mice. *Food Sci. Technol. Res.* **2011,** *17* (2), 93–102.

82. Song, I.; Muller, C.; Louw, J.; Bouwens, L. Regulating the Beta Cell Mass As A Strategy For Type-2 Diabetes Treatment. *Curr. Drug Targets* **2015,** *16* (5), 516–524.

83. Tiwari, B. K.; Pandey, K. B.; Abidi, A. B.; Rizvi, S. I. Therapeutic Potential Of Indian Medicinal Plants In Diabetic Condition. *Ann. Phytomed.* **2013,** *2*, 37–43.

84. Tiwari, B. K.; Pandey, K. B.; Jaiswal, N.; Abidi, A. B.; Rizvi, S. I. Anti-Diabetic And Anti-Oxidative Effect Of Composite Extract Of Leaves Of Some Indian Plants On Alloxan Induced Diabetic Wistar Rats. *J. Pharmaceut. Invest.* **2014,** *44* (3), 205–211.

85. Tomas-Barberen, F. A.; Clifford, M. N. Flavanones, Chalcones And Dihydrochalcones-Nature, Occurrence And Dietary Burden. *J. Sci. Food Agric.* **2000,** *80*, 1073–1080.

86. Torronen, R.; Sarkkinen, E.; Tapola, N.; Hautaniemi, E.; Kilpi, K.; Niskanen, L. Berries Modify The Postprandial Plasma Glucose Response To Sucrose In Healthy Subjects. *Br. J. Nutr.* **2010,** *103* (8), 1094–1097.

87. Unwin, N.; Whiting, D.; Gan, D.; Jacqmain, O.; Ghyoot, G., Eds. IDF Diabetes Atlas, 4th ed.; International Diabetes Federation; Brussels, 2009; p 104.

88. Upadhyay, S.; Dixit, M. Role of Polyphenols and Other Phytochemicals on Molecular Signaling. *Oxid. Med. Cell. Long.* **2015,** *2015*, Article ID: 504253, p 15.

89. Weickert, M. O.; Mohlig, M.; Schofl, C.; Arafat, A. M.; Otto, B.; Viehoff, H. Cereal Fiber Improves Whole Body Insulin Sensitivity In Overweight And Obese Women. *Diabetes Care* **2006,** *29* (4), 775–780.

90. Willett, W. C. Balancing Life-Style And Genomics Research For Disease Prevention. *Science* **2002,** *296*, 695–698.

91. Williamson, G. Possible Effects Of Dietary Polyphenols On Sugar Absorption And Digestion. *Mol. Nutr. Food Res.* **2013,** *57* (1), 48–57.

92. Witkowska, A. M.; Waśkiewicz, A.; Zujko, M. E. Dietary Polyphenol Intake, But Not The Dietary Total Antioxidant Capacity, Is Inversely Related To Cardiovascular Disease In Postmenopausal Polish Women: Results Of WOBASZ And WOBASZ- II Studies. *Oxid. Med. Cell. Longev.* **2017,** *2017*, Article ID: 5982809, 11.

93. Wu, F.; Tang, L.; Chen, B. Oxidative Stress: Implications For The Development Of Diabetic Retinopathy And Antioxidant Therapeutic Perspectives. *Oxid. Med. Cell. Longev.* **2014,** *2014*, Article ID: 752387, 12.

94. Xiao, J. B.; Hogger, P. Dietary Polyphenols And Type 2 Diabetes: Current Insights And Future Perspectives. *Curr. Med. Chem.* **2015**, *22* (1), 23–38.

95. Yagasaki, K. Anti-Diabetic Phytochemicals That Promote GLUT4 Translocation Via AMPK Signaling In Muscle Cells. *Nutr. Aging* **2013/2014**, *2*, 35–44.

96. Youl, E.; Bardy, G.; Magous, R.; Cros, G.; Sejalon, F. Quercetin Potentiates Insulin Secretion And Protects INS-1 Pancreatic B-Cells Against Oxidative Damage Via The ERK1/2 Pathway. *Brit. J. Pharmacol.* **2010**, *161* (4), 799–814.

97. Zang, M.; Xu, S.; Maitland-Toolan, K. A.; Zuccollo, A. Polyphenols Stimulate AMP-Activated Protein Kinase, Lower Lipids, And Inhibit Accelerated Atherosclerosis In Diabetic LDL Receptor–Deficient Mice. *Diabetes* **2006**, *55* (8), 2180–2191.

CHAPTER 8

POTENTIAL OF PHYTOCHEMICALS IN THE TREATMENT OF HEMORRHOIDS

U. KOCA-CALISKAN and C. DONMEZ

ABSTRACT

Hemorrhoid is an anorectal disease that affects more than half of the population in the United States. Majority of the causative factors of hemorrhoid are: sedentary life style, low fiber diet, constipation, obesity, and pregnancy. Hemorrhoid is not a malign perianal illness, and vasodilation on *plexus haemorrhoidalis* vein leads to this disease. Hemorrhoids are usually seen in three areas, which are left lateral, right anterior, and right posterior; and can be categorized as external, internal, and mixed (internal and external). Anti-inflammatory drugs, antibiotics, corticosteroids, local anesthesia, and some plant extracts are used for healing of hemorrhoid. Patients generally choose plant-based intestinal regulators and phlebotonics for general therapy. Plants or plant extracts have been used either directly or with mixed preparation containing some chemicals. There are a limited number of plants to treat hemorrhoids, such as *Aesculus hippocastanum* L., *Hamamelis virginiana* L., *Ruscus aculeatus* L. etc. Phytochemicals to treat hemorrhoids and venous insufficiency are flavonoids, steroidal saponins, terpenes, and triterpenes. This chapter includes brief information on hemorrhoids, anti-hemorrhoidal plants, and most commonly used preparations and their activities.

8.1 INTRODUCTION

Hemorrhoid is an anorectal disease that affects more than half of the population in the United States.[44] It is defined as the flux of blood by Hippocrates.[41] Majority of complaints of the patient include itching, bleeding, pain, and feeling of discomfort. Etiological aspects such as pregnancy, constipation, sedentary lifestyle, obesity, low fiber diet, tight-laced clothes, and seasonal changes might also cause this disease.[44] The hemorrhoidal diseases are celestially categorized considering the anatomical characteristics. The treatment option depends on the degree and severity of symptoms and patient's history.

This chapter includes brief information on hemorrhoids, anti-hemorrhoidal plants, and most commonly used preparations and their activities.

8.1.1 PHYSIOLOGY OF HEMORRHOID

Hemorrhoidal illness is a benign perianal ailment that is actually caused by vasodilation on *plexus haemorrhoidalis* vein. Groups of vascular tissues, smooth muscles, and connective tissues, which are present in all healthy individuals, are also called hemorrhoids technically. They function as cushion, lie down the anal canal primarily in three positions: right anterior, right posterior left lateral, and various numbers of minor cushions lie between them.[64] Hemorrhoids are generally categorized (Fig. 8.1) as external, internal, and interno-external (mixed) according to their position and degree of prolapse.[68] While mucosa covers the internal hemorrhoids, which initiate from the inferior hemorrhoidal venous plexus over the dentate line, squamous epithelium covers the external hemorrhoids, which are dilated veins of this plexus located below the dentate line. As well as, mixed hemorrhoids arise not only above but also below the dentate line. Goligher's classification for internal hemorrhoids is ranked on the basis of their appearance and degree of prolapse[43] as follows:

- Grade I: There is a bleeding on the anal cushions without prolapse.
- Grade II: There is a prolapse on the anal cushions by the anus on straining. It diminishes spontaneously.
- Grade III: It resembles grade II except that it requires manual replacement into the anal canal.
- Grade IV: The anal cushion prolapse stays out of anus every time. Additionally, it is assumed that confined internal hemorrhoid, acutely thrombosed, and thrombosed hemorrhoid bearing peripheral rectal mucosal prolapse are grade IV hemorrhoids.

Painless bright red rectal bleeding is most prevalent expression of hemorrhoids expressed by patients. Prolapsing hemorrhoids may bring about anal itching or perianal irritation due to fecal soiling or/and mucous secretion.

External hemorrhoids and Grade I-II type of internal hemorrhoids can be medicated without surgical operation. Under such circumstances, patients refrain from consulting physicians. This condition leads to a situation so that persons treat themselves via natural practices/products, especially plants and their compounds. Although there is a lot of demand for self-treatment of this disease worldwide, yet the number of plant-based functional pharmaceutical drugs is restricted.[16]

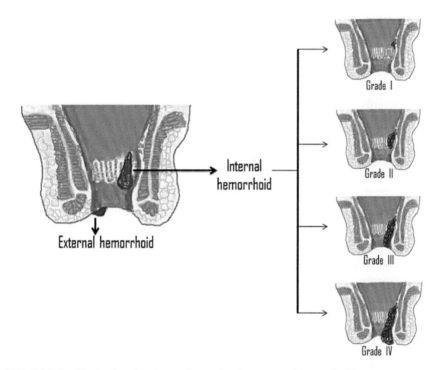

FIGURE 8.1 Illustration showing an internal and an external hemorrhoid.

8.2 TREATMENT OF HEMORRHOIDS

8.2.1 GENERAL TREATMENT OF HEMORRHOID

The treatment options are assessed in a wide spectrum from conventional to surgical procedures. Patient with hemorrhoids requires a careful history and physical, including digital, examination.

The principal objective of the treatment is, in acute phase, to diminish the symptoms rapidly and prevention of the relapses in chronical phase. Patient very often performs local topical treatments, and anti-inflammatory drugs to ameliorate ailment or obstruct the worsening of the hemorrhoid depending on the grade of the ailment, before opting for a surgery. These might include various options, such as use of antiseptics, local anesthesia, anti-inflammatories, vasoconstrictors, antibiotics, corticosteroids, and plant extracts.[33] Synthetic constituents are also mostly preferred, such as zinc oxide and bismuth subgallate as antiseptics and constriction of vessels (or as astringents). Moreover, zinc oxide creates a barrier on the skin to prevent irritation. Cortisone compounds provide relief from anal irritation and itching. Local anesthetics like lidocaine, benzocaine, and cinchocaine chlorhydrate are remedies that cause reversible absence of pain sensation. Fluocortolone is a glucocorticoid with anti-inflammatory activity. Phenylephrine is a decongestant with potent vasoconstrictor property that shrinks blood vessels. Nifedipine is a selective calcium channel blocker that affects blood flow in the vessel and muscle dilation. Nitroglycerin decreases the smooth muscle tonus of blood vessels. Mineral oil, starch, white petrolatum, glycerin, and shark liver oil are also used as protectants.[25,56]. These drugs are effective in the healing of external and first-degree hemorrhoids.

In therapy, drugs are primarily utilized, such as pharmaceutical ointments and creams, oral flavonoids (phlebotonics) tablets, and topical treatments from plant-based fibers. When the therapy is insufficient, surgical procedure is commonly preferred for the treatment. Band ligation method is applied to put a rubber band around the base of the internal hemorrhoids to cease their blood circulation, following the hemorrhoid withers away in a few days. Endoscopic method is used in treatment by destroying the hemorrhoid. Sclerotherapy syringing uses a chemical around the blood vessels to reduce the size of the hemorrhoidal tissue. The last option is hemorrhoidectomy.[78]

8.2.2 ROLE OF NATURAL PHYTOCHEMICALS IN THE TREATMENT OF HEMORRHOIDS

Besides general treatment, plant-based intestinal regulators and plant-based phlebotonics are usually preferred by patients. The astringent effect of plants is parallel to the anti-hemorrhoidal influence.[8] Tannin is a secondary metabolite, which has astringent activity. It is one of the major chemicals of *Hamamelis virginiana* L., *Juglans regia* L., *Cassia* sp., and *Quercus* sp. There are many herbal remedies in the market, containing mostly *Hamamelis virginina* L.[47] Besides, anti-inflammatory activity is closely related to capillary permeability related illness. Flavonoid derivatives are also effective for venous diseases such as varicose vein and hemorrhoid.[61] Additionally, saponin derivatives isolated from *Aesculus hippocastanum* L. and *Ruscus aculeatus* L. are active ingredients in related herbal medicines.

8.2.2.1 EXTRACTS

Aesculus hippocastanum L., *Hamamelis virginiana* L., and *Ruscus aculeatus* L. are the most common plant species in anti-hemorrhoidal preparations as either standardized plant extracts or dried plants and their oils. *Aesculus hippocastanum* L. is standardized on aescin and esculoside; *Ruscus aculeatus* L. is standardized on ruscogenin; and *Hamamelis virginiana* L. is standardized on pyrogallol.[50,65] It was also thought to create a synergistic effect by adding different plants to support the treatment.

8.2.2.2 COMPOUNDS

In Martindale, *The Complete Drug Reference Book*, esculoside, aescin, camphor, diosmin, hesperidin, hesperidin methyl chalcone, menthol, rutoside, tannic acid, thymol, and troxerutin have been recorded as herbal drugs to treat hemorrhoids and venous inadequacy.[62] This reference book may be a guide for phytochemical studies and to discover unfamiliar potent compounds or preparations to cure hemorrhoid or related diseases. Therapeutic effects of the plant extracts and the isolated major substances from extracts can be compared in advanced studies. Discovery and isolation of new/effective compounds, which are flavonoids (hesperidin, diosmin, and troxerutin), volatile or terpenic compounds (camphor, carvacrol, and menthol), saponins (aescin and esculoside), and newly isolated derivatives of these secondary metabolites or other effective compounds will provide contribution to both scientific knowledge and preparation commercial pharmaceutical forms for the remedy of hemorrhoidal diseases (Tables 8.1 and 8.2).

TABLE 8.1 Plant-Based Anti-hemorrhoidal Preparations, Their major Phytochemicals and Their Effective Content (Extract/Compounds), and the Name of the Preparation.

Latin name	Major Phytochemicals	Effective extract or compound for preparations	The name of the preparation	References
Achillea sp.	Flavonoids, Phenolic acids, Essential oils	Extract	Hemoral Slovakofarma®, Natusor Circusil Soria Natural®	[13, 62]
Aesculus hippocastanum L.	Triterpenoids, Triterpenoid Glycosides (Triterpenoid Saponins), Flavonoids, Coumarins, Carotenoids, Long fatty chain Compounds	Extract Coumarin glucoside (Esculoside) Triterpene saponins (Aescin)	Aescorin N Steigerwald®, Aescusan®, Posti N Kade®, Phlebosedol Lehning®, Almodin Nordin®, Aphloine PDB®, Fluon®, Phlebosedol®, Ruscimel Soria Natural®, Supositorio Hamamelis Composto Simoes®, Suppositoires contre les hemorroides Weleda®, Varicare Andromaco®, Varicylum-S Liebermann®, Simoes®, Venoful Garden House®, VNS 45 Sigma®, Reparil®, Hemorrogel Arkopharma®	[58, 62, 76]
Agrimonia eupatoria L.	Carbohydrates, Tannins, Terpenoids, Phenolic compounds (Flavonoids), Agrimony lactone, Glycosides, Volatile oils	Extract	Hemoral Slovakofarma®, Piletabs Potter's®	[4, 62]
Alchemilla sp.	Phenolics compounds, Condensed tannins	Extract	Phlebosedol®	[7, 62]
Alisma sp.	Triterpenoids, Sesquiterpenoids	Extract	TJ-23 Tsumura Toki-Shakuyaku-San Tsumura®	[48, 62]
Allium sativum L.	Amino acids, Anthocyanins, Flavonoids, Essential oil, Steroidal glycosides, Lectins, Pectin, Fructan, Prostaglandins, Vitamins, Lipids	Extract	Ajomast Circulatorio Monserrat®	[15, 62]

TABLE 8.1 *(Continued)*

Latin name	Major Phytochemicals	Effective extract or compound for preparations	The name of the preparation	References
Aloe vera (L.) Burm. f.	Monosaccharides, Amino acids, Vitamins, Minerals	Extract	Flucirac®, Neo-Healar Arab®	[23, 62]
Angelica sp.	Angelicide, Brefeldin A, Adenine, Uracil, Butylidenephthalide, Butyphthalide, Succinic acid, Ferulic acid, Ligustilide, Nicotinic acid	Extract	TJ-23 Tsumura Toki-Shakuyaku-San Tsumura®	[62, 77]
Arnica montana L.	Essential oil, Triterpenols, Sterols, Carotenoides	Extract	Arnikamill Biomo®, Varicylum-S Liebermann®	[9, 62]
Asperula odorata L.	Anthraquinones, Essential oils, Phenol carboxylic acids, Coumarins, Flavonoids, Tannins, Saponins	Extract	Bionorica®	[54, 62]
Atractylodes lancea Thunb.	Polyacetylenes, Sesquiterpenoids, and Sesquiterpene Glycosides	Extract	TJ-23 Tsumura Toki-Shakuyaku-San Tsumura®	[62, 71]
Atropa belladonna L.	Alkaloids, Flavonoids, Chlorogenic acid	Extract	Antiemorroidali AFOM®, Hemosedan Centrapharm®	[17, 62]
Calendula officinalis L.	Coumarines, Carotenoids, Flavonoids, Triterpenoids, Volatile oil, Quinones, Amino acids	Extract	Flucirac®, Hemorrogel Arkopharma®	[46, 62]
Capsella bursa-pastoris (L.) Medik.	Flavonoids, Polypeptides, Choline, Acetylcholine, Histamine, Tyramine, Minerals, Vitamins, Fatty acids	Extract	Proctosor Soria Natural®	[3, 62]

TABLE 8.1 (Continued)

Latin name	Major Phytochemicals	Effective extract or compound for preparations	The name of the preparation	References
Cassia sp.	Alkaloids, Tannins, Saponins, Anthraquinones, Anthocyanosides, Flavonoids	Extract	Circanetten Eversil®, Agiolax Phoenix®	[62, 69]
Centella asiatica (L.) Urb.	Flavonoids, Tocopherols, Phosphatidic and other organic acids, Derivatives of cinnamic acid	Extract Triterpene (Asiatic acid, Madecassic acid)	Flebovis®, Madecassol®, Pertusan®, Venoful Garden House®, Venos Soho®, VNS 45 Sigma®	[62, 74]
Cinnamomum camphora(L.) Presl.	Alkaloids, Essential oils	Terpenoid derivative (Camphor)	Esculeol P®, Manzan Gezzi®, Pazo Bristol-Myers Products®, Chefaro®	[40, 62]
Cnidium monnieri (L.) Cusson ex Juss.	Sesquiterpenes, Coumarins, Chromones, Essential oils	Extract	TJ-23 Tsumura Toki-Shakuyaku-San Tsumura®	[62, 75]
Coleus atropurpureus Benth.	Terpenoid compounds, Essential oils, Saponins, Catechins	Extract	Ambeven Medikon®	[1, 62]
Collinsonia canadensis L.	Flavonoids, Terpenoids	Extract	Piletabs Potter's®	[60, 62]
Crataegus sp.	Oligomeric procyanidins, Flavonoids, Triterpenes, Polysaccharides, Catecholamines	Extract	Bionorica®	[39, 62]
Cuminum cyminum L.	Essential oil, Phenolic compounds	Extract	Flebovis®	[14, 62]
Cupressus sp.	Monoterpenes, Sesquiterpenes, Diterpenes	Extract	Natusor Circusil Soria Natural®, Proctosor Soria Natural®, Ruscimel Soria Natural®, Veinostase Merck Medication Familiale®	[38, 62]

TABLE 8.1 (Continued)

Latin name	Major Phytochemicals	Effective extract or compound for preparations	The name of the preparation	References
Curcuma heyneana Valeton & Zijp	Diarylheptanoids, Sesquiterpenoids	Extract	Ambeven Medikon®	[6, 62]
Echinacea sp.	Alkamides, Phenylpropanoids, Polysaccharides, Volatile oils	Extract	Bionorica®	[11, 62]
Erechtites hieraciifolius (L.) Raf. ex DC.	Dietary fibers, Minerals, Vitamin C	Extract	Hemorrogel Arkopharma®, Piletabs Potter's®	[59, 62]
Euphorbia sp.	Cerebrosides, Terpenoids (Di-, Tri-, Sesqui-), Flavonoids, Glycerols, Steroids, Phenolics	Extract	Thankgod Panacea®	[55, 62]
Extractago sp.	Carbohydrates, Lipids, Alkaloids, Vitamins, Caffeic acid derivatives, Flavonoids, Iridoid glycosides, Triterpenoids, Glucosinolates	Extract	Agiolax Phoenix®, Fybogel®, Metamucil®, PC Regulax Procare ®	[52, 62]
Ginkgo biloba L.	Catechins, Carboxylic acids, Flavonol glycosides, Proanthocyanidins, Terpene trilactones	Extract	Ginkor Fort®, Perivar Ipsen®, Venoful Garden House®, Veno-Tebonin N Schwabe®, VNS 45 Sigma®	[62, 66]
Graptophyllum pictum (L.) Griff.	Alkaloid, Pectin, Formic Acid, Steroid, Saponin, Tannin, Flavonoid, Alcohol, Vomifoliol, Flavonoids, Essential oils	Extract	Ambeven Plus Medikon®, Ambeven Medikon®, Papaven Medikon®	[32, 62, 70]

TABLE 8.1 (Continued)

Latin name	Major Phytochemicals	Effective extract or compound for preparations	The name of the preparation	References
Hamamelis virginiana L.	Tannins, Anthocyanidins, Volatile oil, Flavonoids, Saponins, Resin, Wax	Extract	Hametan®, Almodin Nordin®, Aphloine P DB®, Fluon®, Phlebosedol®, Ruscimel Soria Natural®, Mirorroidin Sedabel®, Flucirac®, Fluon®, Heemex Lane®, Hemodotti Prodotti®, Onrectal Herbes Universelles®, Phlebosedol Lehning®, Rectovasol Qualiphar®, Similia Hochstetter®, Supositorio Hamamelis Composto Simoes®, Suppositoires contre les hemorroides Weleda®, Varicare Andromaco®, Varicylum-S Liebermann®, Simoes®, Venofuil Garden House®, VNS 45 Sigma®	[21, 62]
Hydrastis canadensis L.	Isoquinoline alkaloids, Quinic acid feruloyl esters	Extract	Proctosor Soria Natural®	[27, 62]
Juglans regia L.	Protein, Dietary fiber, Melatonin, Extract sterols, Folate, Tannins, Polyphenols	Extract	Natusor Circusil Soria Natural®	[49, 62]
Kaempferia angustifolia Roscoe	Cyclohexane diepoxide derivatives, Flavonoids, Diterpenes	Extract	Ambeven Medikon®	[62, 63]
Krameria sp.	Lignans, Caffeoylesters, Catechin-glucosides, Dimeric procyanidines, Phenylpropanoyd derivatives	Extract	Esculeol P®, Hemorrol Hochstetter®	[24, 62]
Lupinus albus L.	Quinolizidine alkaloids, Protein, Fiber, Oil, Sugar	Extract	Neo-Healar Arab®	[22, 62]

TABLE 8.1 (Continued)

Latin name	Major Phytochemicals	Effective extract or compound for preparations	The name of the preparation	References
Malva sp.	Ascorbic acid, Carotenoids, Fatty acids, Flavonoids, Sugars, Tocopherols, Phenolics	Extract	Mictasol Bleu Formenti®	[12, 62]
Matricaria chamomilla L.	Flavonoids, Mucilages, Phenyl carbonic acids, Amino acids	Extract Oil Guaiazulene	Arnikamill Biomo®, Hedensa®, Hemoral Slovakofarma®, Varicylum-S Liebermann®, Thrombocid®	[57, 62]
Melilotus officinalis (L.) Lam.	Coumarin glycosides, Essential oils, Flavonoid	Extract	Pierre Fabre®, Esberiven Fort®, Flebovis®, Meli Rephastasan Repha®	[5, 62]
Mentha sp.	Essential oils	Extract Terpene (Menthol)	Neo-Healar Arab®, Hemoral Slovakofarma®, Flavovenyl®, Hamos N®, Hedensa®, Hemoralgine®, Mirorroidin Sedabel®, Procto Venart Casasco®, Proctosan®, Tewa®, Rowatanal®	[62, 26]
Myroxylon sp.	Cinnamic acid, Benzoic acid and esters, Vanillin, Terpenic resins	Resin (Peru balsam)	Anusol®, Rectovasol Qualiphar®	[36, 62]
Paeonia sp.	Monoterpenes and monoterpene glycosides, Triterpenoids, Flavonoids, Phenols, Tannins	Extract	TJ-23 Tsumura Toki-Shakuyaku-San Tsumura®	[62, 73]
Phyllanthus niruri L.	Alkaloids, Coumarins, Flavonoids, Lignans, Polyphenols, Tannins, Terpenoids, Saponins	Extract	Flucirac®	[10, 62]

TABLE 8.1 *(Continued)*

Latin name	Major Phytochemicals	Effective extract or compound for preparations	The name of the preparation	References
Quercus sp.	Tannins, Flavonoids, Alkaloids, Saponins, Cardiac glycosides, Steroids	Extract	Dubova Kura®, Hemoral Slovakofarma®	[53, 62]
Rhamnus purshiana DC.	Anthraquinones	Extract	Modern Herbals Pile®, Pileabs Lane®, Piletabs Potter's®	[62, 67]
Ribes nigrum L.	Anthocyanins, Proanthocyanidins, Quercetin, Myricetin, Phenolic acids, Isorhamnetin, Vitamins	Extract	Veinobiase Fournier®	[35, 62]
Rubia cordifolia L.	Anthraquinones, Prenyl naptho-quinones, Triterpenes	Extract	Ambeven Medikon®	[30, 62]
Ruscus aculeatus L.	Steroidal saponins, and their sulfated–acetylated derivates, Coumarins, Flavonoids	Extract Steroidal Saponins (Ruscogenin)	Cirkan Pierre Fabre®, CVP Flebo Phoenix®, Fagorutin Ruscus®, Miopropan Proctologico Bernabo®, Phlebodril mono Pierre Fabre®, Robapharm®, Ruscimel Soria Natural®, Venyl Dolisos®, Venos Soho®, Venobiase mono Fournier®, Venidium Ivax®, Miopropan Proctologico Bernabo®, Procto Venart Casasco®, Proctolog®	[20, 62]
Ruta graveolens L.	Coumarins, Alkaloids, Terpenes, Flavonoids	Extract	Bionorica®	[37, 62]
Sanguisorba obtusa Maxim.	Phenolic compounds, Tannins, Saponin	Extract	Ambeven Medikon®	[28, 62]
Silybum marianum (L.) Gaertn.	Flavonolignan	Extract	Bionorica®	[19, 62]

TABLE 8.1 (Continued)

Latin name	Major Phytochemicals	Effective extract or compound for preparations	The name of the preparation	References
Soja hispida/ Glycine max	Isoflavones Proteins, Dietary fiber	Oil	Flebovis®	[51, 62]
Sophora japonica L.	Alkaloids, Phospholipids, Polysaccharides, Triterpenes, Amino acids, Fatty acids	Extract	Ambeven Medikon®	[62, 72]
Triticum sp.	Flavonoids, Lignans	Extract	Trifyba Sanofi Synthelabo®	[18, 62]
Ulmus sp.	Lignan, Triterpene esters, Catechin, Catechin rhamnoside, Catechin apiofuranoside, Neolignan glycosides, Sesquiterpene O-naphthaquinones	Extract	Modern Herbals Pile®, PC Regulax Procare®, Pileabs Lane®, Piletabs Potter's®	[34, 62]
Vaccinium myrtillus L.	Anthocyanins, Proanthocyanidins Chlorogenic acid derivatives, Hydroxycinnamic acids, Flavonol glycosides, Catechins	Extract	Myrtaven IBSA®, Varison®	[45, 62]
Vateria indica L.	Stilbenes, Flavonols, Diterpenes, Isocoumarin, Phenylpropanoid	Extract	Neo-Healar Arab®	[29, 62]
Viburnum sp.	Diterpenoid, Iridoid, Coumarin, Flavonoids	Extract	Aphloine P DB®, Fluon® , Phlebosedol®	[42, 62]
Vitis vinifera L.	Phenolic compounds, Catechins, Procyanidins	Extract	Jouvence de l'Abbe Soury-Chefaro Ardeval®, Phlebosedol Lehning®, Veinophytum Arkomedika®	[31, 62]
Zingiber officinale Roscoe	Volatile oils, Diarylheptanoids	Extract	PC Regulax Procare®	[2, 62]

TABLE 8.2   The Plant–Based Secondary Metabolites and the Name of Their Preparations.

Effective compounds	Preparation	Reference
Alkaloid derivatives (Ephedrine sulfate, Ephedrinehydrochloride, Cinchocaine)	Pazo Bristol-Myers Products®, Thunas Pile Thuna®, Hemoralgine®, Nupercainal®, Proctosedyl®, Proctyl Altana®	[62]
Coumarin	Microsuy Bernabo®, Venalot Nycomed®	[62]
Flavonoids (Bioflavonoids, Citroflavonoids, Diosmin, Hesperidin, Rutoside)	Flebotropin®, Accesum Roux-Ocefa®,Ardium Servier ®, Ciflon Lapi®, Daflon®, Diosper Kleva®, Esculeol P®, Quali-Flon Quality®, Terbenol Duo Bernabo®, Variton Hormona®, Vedium Medikon®, Vesnidan Medipharm®, Microsuy Bernabo®,Venoruton®	[62]
Flavonol (Troxerutin)	Ambeven Plus Medikon®, Cilkanol Zentiva®, Flavol®, Ginkor Fort®, Jatamansin®, Lindigoa S Riemser®, Microsuy Bernabo®,Veinamitol Negma®, Venalot Nycomed®,VNS 45 Sigma®	[62]
Tannic acid	HEC®, Teva®, Tannolil Li-il®, Chefaro®	[62]
Terpene (Monochlorocarvacrol, Thymol)	Hedensa®, Epianal®, Manzan Gezzi®, Teva®	[62]

8.3   SUMMARY

In anti-hemorrhoidal preparation, plant extracts are preferred commonly. This study concluded that 61 different plants are medicinal applied directly or indirectly in hemorrhoids. While 58 plants were used as extracts in medicines, 8 different compounds (esculoside, aescin, ruscogenin, guaiazulene, menthol, asiatic acid, madecassic acid, and camphor) obtained from these plants were included in formulations as active ingredients. Not only coumarin, terpenic, steroidal, and triterpenic saponins obtained from these plants but also resin, oil, and terpenic compounds are used directly in pharmaceutical drugs. Alkaloid derivatives (ephedrine sulfate, ephedrinehydrochloride, and cinchocaine), coumarin, flavonoids (hesperidin, diosmin, bioflavonoids, citroflavonoids, rutoside, and troxerutin), tannic acid, and terpenic (Thymol and monochlorocarvacrol) compounds are commonly found in hemorrhoid medications. The five plants (*Aesculus hippocastanum* L., *Centella asiatica* (L.) Urb., *Matricaria chamomilla* L., *Mentha sp.,* and *Ruscus aculeatus* L.) were included in preparations, both as an extract or as an individual phytochemical content.

KEYWORDS

- anti-hemorrhoidal phytochemical
- hemorrhoid
- herbal medicine
- secondary metabolite

REFERENCES

1. Ahmad, A.; Massi, M. N. The Antituberculosis Drug Rifampicin Is Activated By 2', 5'-Dimethyl Benzopelargonolactone From Leaves Of *Coleus Atropurpureus* L. Benth. *Int. J. Pharma Bio Sci.* **2014,** *5* (1), 758–764.

2. Ali, B. H.; Blunden, G.; Tanira, M. O.; Nemmar, A. Phytochemical, Pharmacological And Toxicological Properties Of Ginger (*Zingiber Officinale* Roscoe): A Review Of Recent Research. *Food Chem. Toxicol.* **2008,** *46* (2), 409–420.

3. Al-Snafi, A. E. The Chemical Constituents And Pharmacological Effects Of *Capsella Bursa-Pastoris*-A Review. *Int. J. Pharmacol. Toxicol.* **2015,** *5* (2), 76–81.

4. Al-Snafi, A. E. The Pharmacological And Therapeutic Importance Of *Agrimonia Eupatoria* -A Review. *Asian J. Pharm. Sci. Technol.* **2015,** *5* (2), 112–117.

5. Anwer, M. S.; Mohtasheem, M.; Azhar, I.; Ahmed, S. W.; Bano, H. Chemical Constituents From *Melilotus Officinalis*. *J. Basic Appl. Sci.* **2008,** *4* (2), 8994.

6. Aspollah Sukari, M.; Wah, T. S.; Saad, S. M.; Rashid, N. Y.; Rahmani, M.; Lajis, N. H.; Hin, T. Y. Y. Bioactive Sesquiterpenes From *Curcuma Ochrorhiza* And *Curcuma Heyneana*. *Nat. Prod. Res.* **2010,** *24* (9), 838–845.

7. Ayaz, A. F.; Hayirlioglu-Ayaz, S. Total Phenols And Condensed Tanins In The Leaves Of Some *Alchemilla* Species. *Sect. Cell. Mol. Biol.* **2001,** *56* (4), 449–453.

8. Azeemuddin, M.; Viswanatha, G. L.; Rafiq, M. An Improved Experimental Model Of Hemorrhoids In Rats: Evaluation Of Antihemorrhoidal Activity Of An Herbal Formulation. *ISRN 530931 Pharmacol.* **2014,** *2014*, 1–7.

9. Azimova, S. S.; Glushenkova, A. I. *Arnica Montana* L. In *Lipids, Lipophilic Components and Essential Oils from Plant Sources*; Springer: London, 2012; pp 55–56.

10. Bagalkotkar, G.; Sagineedu, S. R.; Saad, M. S.; Stanslas, J. Phytochemicals From *Phyllanthus Niruri* Linn. And Their Pharmacological Properties: A Review. *J. Pharm. Pharmacol.* **2006,** *58* (12), 1559–1570.

11. Barnes, J.; Anderson, L. A.; Gibbons, S.; Phillipson, J. D. Echinacea Species (*Echinacea Angustifolia* (DC.) Hell.; *Echinacea Pallida* (Nutt.) Nutt.; *Echinacea Purpurea* (L.) Moench): Review Of Their Chemistry, Pharmacology And Clinical Properties. *J. Pharm. Pharmacol.* **2005,** *57* (8), 929–954.

12. Barros, L.; Carvalho, A. M.; Ferreira, I. C. Leaves, Flowers, Immature Fruits And Leafy Flowered Stems Of *Malva Sylvestris*: A Comparative Study Of The Nutraceutical Potential And Composition. *Food Chem. Toxicol.* **2010,** *48* (6), 1466–1472.

13. Benedek, B.; Kopp, B. *Achillea millefolium* L. Revisited: Recent Findings To Confirm The Traditional Use. *Wien. Med. Wochenschr.* **2007**, *157* (13–14), 312–314.

14. Bettaieb, I.; Bourgou, S.; Wannes, W. A.; Hamrouni, I.; Limam, F.; Marzouk, B. Essential Oils, Phenolics, And Antioxidant Activities Of Different Parts Of Cumin (*Cuminum Cyminum* L.). *J. Agric. Food Chem.* **2010**, *58* (19), 10410–10418.

15. Bozin, B.; Mimica-Dukic, N.; Samojlik, I.; Goran, A. Phenolics as Antioxidants In Garlic (*Allium Sativum* L., Alliaceae). *Food Chem.* **2008**, *111* (4), 925–929.

16. Caliskan, U. K.; Aka, C.; Oz, M. G. Plants Used In Anatolian Traditional Medicine For The Treatment Of Hemorrhoid. *Rec. Nat. Prod.* **2017**, *11* (3), 235–250.

17. Capasso, R.; Izzo, A. A.; Pinto, L.; Bifulco, T.; Vitobello, C.; Mascolo, N. Phytotherapy And Quality Of Herbal Medicines. *Fitoterapia* **2000**, *71*, 58–65.

18. Cooper, R. Re-Discovering Ancient Wheat Varieties As Functional Foods. *J. Tradit. Complement. Med.* **2015**, *5* (3), 138143.

19. Davis-Searles, P. R.; Nakanishi, Y.; Kim, N. C.; Graf, T. N.; Oberlies, N. H.; Wani, M. C.; Kroll, D. J. Milk Thistle And Prostate Cancer: Differential Effects Of Pure Flavonolignans From *Silybum Marianum* On Antiproliferative End Points In Human Prostate Carcinoma Cells. *Cancer Res.* **2005**, *65* (10), 4448–4457.

20. Edwards, S. E.; Rocha, I. D. C.; Williamson, E. M.; Heinrich, M. Butcher's Broom *Ruscus aculeatus* L. In *Phytopharmacy: An Evidence-Based Guide to Herbal Medicinal Products*; Wiley-Blackwell Press: New York:, 2015; pp 75100.

21. Edwards, S. E.; Rocha, I. D. C.; Williamson, E. M.; Heinrich, M. Witch Hazel. In *Phytopharmacy: An Evidence-Based Guide to Herbal Medical Products*; Wiley-Blackwell Press: New York, 2015; pp 396–400.

22. Erbas, M.; Certel, M.; Uslu, M. K. Some Chemical Properties of White Lupin Seeds (*Lupinus Albus* L.). *Food Chem.* **2005**, *89* (3), 341–345.

23. Eshun, K.; He, Q. *Aloe Vera*: A Valuable Ingredient For The Food, Pharmaceutical And Cosmetic Industries: Review. *Crit. Rev. Food Sci. Nutr.* **2004**, *44* (2), 91–96.

24. Facino, R. M.; Carini, M.; Aldini, G.; De Angelis, L. Rapid Screening By Liquid Chromatography/Mass Spectrometry And Fast Atom Bombardment Tandem Mass Spectrometry Of Phenolic Constituents With Radical Scavenging Activity, From *Krameria Triandra* Roots. *Rapid Commun. Mass. Spectrom.* **1997**, *11* (12), 1303–1308.

25. Friciu, M.; Chefson, A.; Leclair, G. Stability of Hydrocortisone, Nifedipine, And Nitroglycerine Compounded Preparations For The Treatment of Anorectal Conditions. *Can. J. Hosp. Pharm.* **2016**, *69* (4), 329–333.

26. Hussain, A. I.; Anwar, F.; Nigam, P. S.; Ashraf, M.; Gilani, A. H. Seasonal Variation In Content, Chemical Composition And Antimicrobial And Cytotoxic Activities Of Essential Oils From Four Mentha Species. *J. Sci. Food Agric.* **2010**, *90* (11), 1827–1836.

27. Hwang, B. Y.; Roberts, S. K.; Chadwick, L. R.; Wu, C. D.; Kinghorn, A. D. Antimicrobial Constituents From Rhizomes Of Goldenseal (*Hydrastis Canadensis*) Against Selected Oral Pathogens. *Planta Med.* **2003**, *69* (7), 623–627.

28. Ishimaru, K.; Hirose, M.; Takahashi, K.; Koyama, K.; Shimomura, K. *Sanguisorba officinalis* L. (Great Burnet): *In Vitro* Culture And Production Of Sanguiin, Tannins, And Other Secondary Metabolites. In *Medicinal and Aromatic Plants VIII*; Springer Berlin: Heidelberg, Germany, 1995; pp 427–441.

29. Ito, T.; Masuda, Y.; Abe, N.; Oyama, M.; Sawa, R.; Takahashi, Y.; Iinuma, M. Chemical Constituents In The Leaves Of *Vateria indica*. *Chem. Pharm. Bull.* **2010**, *58* (10), 1369–1378.

30. Itokawa, H.; Mihara, K.; Takeya, K. Studies On A Novel Anthraquinone And Its Glycosides Isolated From *Rubia Cordifolia* And *R. Akane. Chem. Pharm. Bull.* **1983,** *31* (7), 2353–2358.

31. Jayaprakasha, G. K.; Selvi, T.; Sakariah, K. K. Antibacterial And Antioxidant Activities Of Grape (*Vitis Vinifera*) Seed Extracts. *Food Res. Int.* **2003,** *36* (2), 117–122.

32. Jiangseubchatveera, N.; Liawruangrath, B.; Liawruangrath, S. The Chemical Constituents And The Cytotoxicity, Antioxidant And Antibacterial Activities Of The Essential Oil Of *Graptophyllum Pictum* (L.) Griff. *J. Essent. Oil Bearl. Pl.* **2015,** *18* (1), 11–17.

33. Johnson, J. F. Nonsurgical Treatment Of Hemorrhoids. *J. Gastrointest. Surg.* **2002,** *6,* 290–294.

34. Jung, M. J.; Heo, S. I.; Wang, M. H. Free Radical Scavenging And Total Phenolic Contents From Methanolic Extracts Of *Ulmus Davidiana. Food Chem.* **2008,** *108* (2), 482487.

35. Karjalainen, R.; Anttonen, M.; Saviranta, N.; Stewart, D. Review On Bioactive Compounds In Black Currants (*Ribes Nigrum* L.) And Their Potential Health-Promoting Properties. *Int. Symp. Biotechnol. Fruit Species: BIOTECHFRUIT2008* **2008,** *839* (September), 301–307.

36. Kaymak, Y.; Tırnaksız, F. *Kozmetik ürünlere bağli istenmeyen etkiler* (Adverse Effects Related to Cosmetic Products). *Dermatose* **2007,** *6* (1), 39–48.

37. Kostova, I.; Ivanova, A.; Mikhova, B.; Klaiber, I. Alkaloids And Coumarins From *Ruta Graveolens. Monats. Chem./Chem. Mon.* **1999,** *130* (5), 703–707.

38. Kuiate, J. R.; Bessiere, J. M.; Zollo, P. H. A.; Kuate, S. P. Chemical Composition And Antidermatophytic Properties Of Volatile Fractions Of Hexanic Extract From Leaves Of *Cupressus Lusitanica* Mill. From Cameroon. *J. Ethnopharmacol.* **2006,** *103* (2), 160–165.

39. Kumar, D.; Arya, V.; Bhat, Z. A.; Khan, N. A.; Prasad, D. N. The Genus Crataegus: Chemical And Pharmacological Perspectives. *Rev. Bras. Farmacogn.* **2012,** *22* (5), 1187–1200.

40. Lee, H. J.; Hyun, E. A.; Yoon, W. J.; Kim, B. H.; Rhee, M. H. *In vitro* Anti-Inflammatory And Anti-Oxidative Effects Of *Cinnamomum Camphora* Extracts. *J. Ethnopharmacol.* **2006,** *103* (2), 208216.

41. Leff, E. Hemorrhoids: Current Approaches To An Ancient Problem. *Postgrad. Med.* **1987,** *82* (7), 95–101.

42. Lobstein, A.; Haan-Archipoff, G.; Englert, J. Chemotaxonomical Investigation In The Genus *Viburnum. Phytochemistry* **1999,** *50* (7), 1175–1180.

43. Lunniss, P. J.; Mann, C.V. Classification Of Internal Haemorrhoids: A Discussion Paper. *Colorectal. Dis.* **2004,** *6,* 226–232.

44. MacKay, D. Hemorrhoids And Varicose Veins: A Review Of Treatment Options. *Altern. Med. Rev.* **2001,** *6* (2), 126–126.

45. Martz, F.; Jaakola, L.; Julkunen-Tiitto, R.; Stark, S. Phenolic Composition And Antioxidant Capacity Of Bilberry (*Vaccinium Myrtillus*) Leaves In Northern Europe Following Foliar Development And Along Environmental Gradients. *J. Chem. Ecol.* **2010,** *36* (9), 1017–1028.

46. Muley, B. P.; Khadabadi, S. S.; Banarase, N. B. Phytochemical Constituents And Pharmacological Activities Of *Calendula Officinalis* Linn (Asteraceae): A Review. *Trop. J. Pharm. Res.* **2009,** *8* (5), 455–465.

47. Odukoya, O. A.; Sofidiya, M. O.; Ilori, O. O.; Gbededo, M. O. Hemorrhoid Therapy With Medicinal Plants: Astringency And Inhibition Of Lipid Peroxidation As Key Factors. *J. Biol. Chem.* **2009,** *3* (3), 111–118.

48. Peng, G. P.; Tian, G.; Huang, X. F.; Lou, F. C. Guaiane-Type Sesquiterpenoids From *Alisma Orientalis. Phytochemistry* **2003,** *63* (8), 877881.

49. Pereira, J. A.; Oliveira, I.; Sousa, A.; Ferreira, I. C.; Bento, A. Bioactive Properties And Chemical Composition Of Six Walnut (*Juglans Regia* L.) Cultivars. *Food Chem. Toxicol.* **2008,** *46* (6), 2103–2111.

50. Ph.Eur (European Pharmacopoeia) Hamamelis leaf. Monograph: 04/2008:0909; **2007;** 6.1: 3471.

51. Redondo-Cuenca, A.; Villanueva-Suárez, M. J.; Rodríguez-Sevilla, M. D.; *Mateos-Aparicio,* I. Chemical Composition And Dietary Fiber Of Yellow And Green Commercial Soybeans (*Glycine Max*). *Food Chem.* **2007,** *101* (3), 1216–1222.

52. Samuelsen, A. B. The Traditional Uses, Chemical Constituents And Biological Activities Of *Plantago* Major L. - Review. *J. Ethnopharmacol.* **2000,** *71* (1), 1–21.

53. Sánchez-Burgos, J. A.; Ramírez-Mares, M. V.; Larrosa, M. M.; Gallegos-Infante, J. A.; González-Laredo, R. F.; Medina-Torres, L. Antioxidant, Antimicrobial, Antitopoisomerase And Gastroprotective Effect Of Herbal Infusions From Four *Quercus* Species. *Ind. Crops Prod.* **2013,** *42*, 57–62.

54. Sergeevna, I. N.; Mihaylovna, K. A.; Leonidivna, T. E.; Aleksandrovna, K. I. The Antihypoxic And Sedative Activity Of The Dry Extract From *Asperula Odorata* L. *Phcog. Commn.* **2015,** *5* (4), 233–236.

55. Shi, Q. W.; Su, X. H.; Kiyota, H. Chemical And Pharmacological Research Of The Plants In Genus *Euphorbia. Chem. Rev.* **2008,** *108* (10), 4295–4327.

56. Singh, A.; Jain, R.; Singla, A. U.S. Patent 5,858,371. Washington, DC: U.S. Patent and Trademark Office; 1999; p 31.

57. Singh, O.; Khanam, Z.; Misra, N., Srivastava, M. K. Chamomile (*Matricaria Chamomilla* L.): An Overview. *Pharmacogn. Rev.* **2011,** *5* (9), 82–85.

58. Sirtori, C. R. Aescin: Pharmacology, Pharmacokinetics And Therapeutic Profile. *Pharmacol. Res.* **2001,** *44* (3), 183–193.

59. Srianta, I.; Patria, H. D.; Arisasmita, J. H.; Epriliati, I. Ethnobotany, Nutritional Composition And DPPH Radical Scavenging Of Leafy Vegetables Of Wild *Paederia Foetida* And *Erechtites Hieracifoli*a. *Int. Food Res. J.* **2012,** *19* (1), 245–250.

60. Stevens, J. F.; Ivancic, M.; Deinzer, M. L.; Wollenweber, E. Novel 2-Hydroxyflavanone From *Collinsonia Canadensis. J. Nat. Prod.* **1999,** *62* (2), 392–394.

61. Struckmann, J. R.; Nicolaides, A. N. Flavonoids - Review Of The Pharmacology And Therapeutic Efficacy Of Daflon 500 Mg In Patients With Chronic Venous Insufficiency And Related Disorders. *Angiology* **1994,** *45*, 419428.

62. Sweetman, S. *Martindale-The Complete Drug Reference*; 36th ed.; Pharmaceutical Press: London, UK, 2009; pp 114–118.

63. Tang, S. W.; Sukari, M. A.; Neoh, B. K.; Yeap, Y. S. Y. Phytochemicals From *Kaempferia Angustifolia* Rosc. and Their Cytotoxic and Antimicrobial Activities. *BioMed Res. Int.* **2014,** *2014*, 1–6.

64. Thomson, W. H. The Nature Of Haemorrhoids. *Br. J. Surg.* **1975,** *62*, 542–552.

65. Ustunes, L. Hemoroids. *Rx Media Pharma®,* https://www.rxmediapharma.com (accessed Nov 9, 2016); p 8.

66. Van-Beek, T. A. Chemical analysis Of *Ginkgo* Biloba Leaves And Extracts. *J. Chromatogr. A* **2002,** *967* (1), 21–55.

67. Vanden-Berg, A. J. J.; Labadie, R. P. Anthraquinones, Anthrones And Dianthrones In Callus Cultures Of *Rhamnus Frangula* And *Rhamnus Purshiana*. *Planta Med.* **1984,** *50* (5), 449–451.

68. Varut, L. Hemorrhoids: From Basic Pathophysiology To Clinical Management. *World J. Gastroenterol.* **2012,** *18* (17), 2009–2017.

69. Veerachari, U.; Bopaiah, A. K. Phytochemical Investigation Of The Ethanol, Methanol And Ethyl Acetate Leaf Extracts Of Six Cassia Species. *Int. J. Pharma Bio. Sci.* **2012,** *3*, 260–270.

70. Wahyuningtyas, E. The *Graptophyllum Pictum* Extract Effect On Acrylic Resin Complete Denture Plaque Growth. *Dental J. Majalah Kedokteran Gigi-DJMKG* **2005,** *38* (4), 201–204.

71. Wang, H. X.; Liu, C. M.; Liu, Q.; Gao, K. Three Types Of Sesquiterpenes From Rhizomes Of *Atractylodes Lancea*. *Phytochemistry* **2008,** *69* (10), 2088–2094.

72. Wang, J. H.; Lou, F. C.; Wang, Y. L.; Tang, Y. P. Flavonol Tetra-Glycoside From *Sophora Japonica* Seeds. *Phytochemistry* **2003,** *63* (4), 463–465.

73. Wu, S. H.; Wu, D. G.; Chen, Y. W. Chemical Constituents And Bioactivities Of Plants From The Genus Paeonia. *Chem. Biodiver.* **2010,** *7* (1), 90104.

74. Zainol, M. K.; Abd-Hamid, A.; Yusof, S.; Muse, R. Antioxidative Activity And Total Phenolic Compounds Of Leaf, Root And Petiole Of Four Accessions Of *Centella Asiatica* (L.) Urban. *Food Chem.* **2003,** *81* (4), 575–581.

75. Zhang, Q.; Qin, L.; He, W.; Van-Puyvelde, L.; Maes, D.; Adams, A.; De Kimpe, N. Coumarins From *Cnidium Monnieri* And Their Antiosteoporotic Activity. *Planta Med.* **2007,** *73* (1), 13–19.

76. Zhang, Z.; Li, S.; Lian, X. Y. An Overview Of Genus *Aesculus* L.: Ethnobotany, Phytochemistry, And Pharmacological Activities. *Pharm. Crops* **2010,** *1*, 24–51.

77. Zhao, K. J.; Dong, T. T.; Tu, P. F.; Song, Z. H.; Lo, C. K.; Tsim, K. W. Molecular Genetic And Chemical Assessment Of *Radix Angelica* (Danggui) In China. *J. Agric. Food Chem.* **2003,** *51* (9), 2576–2583.

78. Zhifei, S., John, M. Review of Hemorrhoid Disease: Presentation And Management. *Clin. Colon Rectal Surg.* **2016,** *29*, 22–29.

CHAPTER 9

ROLE OF PLANT-BASED BIOFLAVONOIDS IN COMBATING TUBERCULOSIS

ALKA PAWAR and YATENDRA KUMAR SATIJA

ABSTRACT

Despite combination therapy for the management of tuberculosis (TB), the disease is still a huge challenge to the society. The emergence of multidrug-resistant strains has now threatened the scientific community across the globe and therefore new therapies are urgently needed. Since ancient times, phytochemicals have been the basis of almost all therapeutics and, even today, natural products are still entering clinical trials as well as providing new avenues for biocompounds against TB. Flavonoids are most attractive for anti-mycobacterial drugs because of the rich chemical diversity and antibacterial activity. Furthermore, many fruits and vegetables are rich in flavonoids, whose consumption can be considered as relatively safe. Several flavonoids (such as catechin, epigallocatechin gallate, quercetin, naringenin, etc.) are highly effective against TB.

9.1 INTRODUCTION

Tuberculosis (TB) is a chief cause of morbidity and death worldwide. Five decades of TB-preventive programs, with use of available competent chemotherapy as well as the accessibility of Bacille-Calmette Guerin (BCG) vaccine, have not yielded expected results to decrease the occurrence of TB infection. According to a report by World Health Organization (WHO), TB is still the second major cause of death in the world. Two billion persons have latent TB (LTB) infections due to *Mycobacterium tuberculosis* (MTB), and these persons represent one third of the global population.[82]

Current cure for TB consists of four months of rigorous treatment with first-line drugs (namely: isoniazid, rifampicin, ethambutol, and pyrazinamide), which is tailed by continuation phase of four to five additional months of treatment with second-line drug therapy (namely: isoniazid and rifampicin).[12] The condition has worsened because of rise in drug resistance,[15] HIV coinfection,[45] patient defiance along chemotherapy, and variable effectiveness of BCG vaccine. There is utmost importance to (1) develop new anti-TB drugs, which should have prospects against both active and latent/persistent TB bacteria; (2) to reduce duration of therapy to encourage patient's compliance; and (3) to minimize phenomenon of drug resistance in mycobacterium species.[12,32] Therefore, the search for novel anti-tubercular drugs is an urgent need in order to fade out the menace of drug resistance and to eventually eliminate TB.

Plant-based natural products are gaining importance in this direction, as they can provide an outline for the advancement of novel frame of drugs. Numerous biocompounds have received substantial interest as potential anti-TB agents.[52] Among these, flavonoids are most potential for anti-mycobacterial drugs because of their rich chemical diversity and abundant antibacterial activity. Furthermore, flavonoids are found in most fruits and vegetables, which constitute part of regular diet, therefore their consumption can be relatively safe. Several flavonoids (such as catechin, epigallocatechin gallate, quercetin, naringenin, etc.) are reported to be highly effective against TB.

This book chapter focuses on recent advances in the characterization of flavonoids as anti-mycobacterial agents and potential TB drugs.

9.2 TUBERCULOSIS: WORLDWIDE HEALTH MENACE

Tuberculosis is an extremely contagious infection, which is caused by the bacteria MTB. According to the WHO report (2016), an estimated 10.4 million patients were infected with TB worldwide. Out of which, 1.4 million TB patients died in 2015, and 1.17 million cases of HIV-associated TB occurred in 2015 out of which 0.39 million of patients died. It is documented that over 3 billion people have already been immunized with BCG vaccine. Despite this, TB infects and slays down beyond 50,000 persons weekly. About one third of the global population is asymptomatically diseased with MTB.[82]

Tuberculosis is an airborne infection that spreads via intake of contaminated droplets. The prime location of infection is the lung, which is called as pulmonary TB (PTB). However, the infection can also occur at sites other than lungs, which is called as extrapulmonary TB (EPTB).[88] The extremity of the infection, that is, latent or active TB, depends principally on the condition of the host's immunity. Active PTB presents symptoms, such as continuous productive cough called sputum, hemoptysis, and loss of hunger, weight loss, night sweats and fever.[25] The infection of LTB is categorized as excessive threat, because there is an invariable likelihood of LTB rolling back to active TB at a subsequent phase and such strains of TB generally turn into resistant strains. The LTB sufferers also serve as pools for transmission of infection, consequently, it is vital to treat these victims efficiently so that the TB does not progress to active TB.[23]

9.3   TRENDS IN DEVELOPMENT OF ANTI-TB DRUGS

Chemotherapy span of available anti-TB drugs is very lengthy, which again poses a serious public health concern. Another major problematic issue is drug resistance that can span from resistance toward one drug to resistance toward multiple drugs,[75,77] which are termed as multidrug resistant (MDR) TB; extensively drug-resistant (XDR) TB; or totally drug-resistant (TDR) TB.

9.3.1   FOUR FIRST-LINE PIONEER ANTI-TB DRUGS

First-line TB drugs are:[16,63] (1) Isoniazid or isonicotinylhydrazide (INH)—discovered in 1951[63]; (2) Pyrazinamide (PZA: an analogue of nicotin-amide—discovered in 1952; (3) Rifampicin (RIF)—discovered in 1957; and (4) Ethambutol (EMB)—discovered in 1962. Available TB drug treatment regimen includes: (1) **Intensive phase:** combined administration of four first-line anti-TB drugs (INH, PZA, RIF, and EMB for six to nine months). In this phase, all four first-line anti-TB drugs are administered, and (2) **Continuation phase:** both RIF and INH are administered.

Isoniazid (INH) is a prodrug, which needs catalytic activation by the enzyme MTB catalase-peroxidase katG, which leads to the formation of INH-NAD complex. This complex prohibits the nicotinamide adenine dinucleotide (NADH)-dependent enoyl-ACP reductase, which belongs to the fatty acid synthase type II system, a key player in the mycolic acid biosynthetic pathway of MTB.[86] Hence, it inhibits the synthesis of mycolic acid, that is, a cell wall component.

Pyrazinamide (PZA) is an analogue of nicotinamide. It is a prodrug, which requires conversion to pyrazinoic acid by enzyme MTB pyrazinamidase.[85] It possesses an in vivo-enhanced lesion cleansing action and interferes in the production of virulence factors that highlight powerful antibacterial activity. PZA mainly attacks the cell membrane of mycobacteria. Ribosomal protein SA (RpsA), which encodes for ribosomal protein S1 (an essential protein implicated in protein translation as well as in the ribosome-sparing process of trans-translation), has been reported as the cellular target of PZA.[67] Over-expression of RpsA has been correlated with PZA resistance in MTB.

Rifampicin (RIF) binds to beta RNA polymerase of bacteria and inhibits the RNA synthesis. It includes a set of antibacterial drugs and contains various derivatives viz., rifampicin, rifapentine, rifabutin, and rifalazil.[3] Drug resistance is associated with mutation in 81bp region of rpoB gene, which encodes for beta RNA polymerase. INH and RIF with their derivatives are most common used drugs used for TB chemotherapy in combination with additional molecules.

Ethambutol (EMB) mainly targets the synthesis of MTB cell wall by inhibiting polymerization of arabinogalactan, which is an essential constituent of cell wall. Additionally, it also impedes the consumption of the arabinose donor via hindering both the catalytic action of arabinosyl transferase and by forming the arabinose acceptor, or any of them.[81] In MTB, the embCAB operon has demonstrated to be responsible for EMB resistance.[74]

9.3.2   SECOND-LINE ANTI-TB DRUGS

Second-line drugs are: (1) Streptomycin; (2) Para amino salicylic acid; (3) kanamycin; (4) cycloserine; (5) ethionamide; (6) amikacin; (7) capreomycin; (8) fluoroquinolones, and so forth.

Streptomycin was the first antibiotic employed for the tuberculosis therapy and it belongs to aminocyclitol glycoside class. It was originally extracted from *Streptomyces griseus* bacteria. It binds to 16S rRNA and thereby targets the initiation step of protein translation as well as translation proofreading.[4] Mutations in the 16S rRNA and ribosomal protein S12 genes are correlated to the streptomycin resistance in MTB.[70]

Fluoroquinolones were discovered in the year 1965 as a derivative in the purification of the Chloroquine—an antimalarial drug. It mainly inhibits the type II topoisomerase/DNA gyrase enzyme of MTB. The DNA gyrase consists of subunits A and B, coded via gyr genes. Resistance of fluoroqui-nolones mainly occurs due to mutation in gyrA or gyrB genes.[78]

Ethionamide requires activation via mono-oxygenase enzyme, coded by ethA, for the formation of NAD adduct. Hence, it leads to the inhibition of

NADH-dependent enoyl acyl carrier protein reductase. The emergence of ethionamide resistance is correlated with the ethA and inhA gene mutations.[40]

Para-amino salicylate (PAS) is used in a combinatorial approach with isoniazid and streptomycin for anti-TB treatment.[38] Although the mechanism of PAS was never clearly understood, yet it is still used as a successful anti-TB drug. It inhibits dihydropteroate synthase, which is involved in folic acid synthesis. Resistance of PAS occurs due to thymidylate synthase A-gene mutations.[87]

D-cycloserine (DCS) is an analog of D-Alanine, which targets the biosynthesis of the peptidoglycan layer of cell wall by inhibiting d-Alanine: d-alanine ligase. Furthermore, DCS prevents D-Alanine racemase, which is an enzyme essential for the interconversion of L-Alanine and D-Alanine.[20]

Both **kanamycin** and **amikacin** belong to aminoglycosides class of antibiotics. They target mainly the 16S ribosomal subunit and subsequently inhibit the protein biosynthesis. **Capreomycin** and **viomycin** are also inhibitors of cyclic peptide. All these four drugs are employed in the therapy of multidrug resistant TB.[5]

9.3.3   NEW DIRECTIONS IN DISCOVERY OF TB DRUGS

At present, substantial research has been carried out in the advancement of new drugs for the TB treatment. Several lead compounds are already present in the pipeline and some are present in the preclinical stage while others are in clinical trial stage. There are also a number of continuing studies by means of revisited drugs, in which diverse combinations as well as dosages of compounds are in the testing phase for improvement of the TB treatment.

Sirturo (Bedaquiline): In the year 2012, FDA approved a drug called bedaquiline (TMC-207), which is a diarylquinoline anti-mycobacterial drug. It is employed in combination therapy for the cure of adult patients suffering from severe pulmonary multidrug resistant tuberculosis (MDR-TB).[30] It mainly targets the energy metabolism pathway of bacteria by binding to the subunit C of mycobacterial synthase.[19]

Delamanid (OPC67683) and **pretomanid–moxifloxacin–pyrazin-amide** (PA-824) are two latest drugs in combination, which belong to imidazooxazoles. Both drugs are presently in phase III clinical trial. Activation of these pro-drugs depends upon the MTB F420-deazaflavin-dependent nitroreductase (Ddn). The Ddn enzymatically reduces PA-824, which results in intracellular discharge of detrimental reactive nitrogen species that attacks mycobacterial respiration process.[42] The PA-824 drug is active against both active and latent TB, therefore it helps in reducing the TB treatment time.[21] Delamanid drug is used mainly in the cure of multidrug resistant TB and it attacks the biosynthesis of mycolic acid, a component of cell wall. When pooled with other anti-TB drugs, it shows maximum efficiency with tolerable toxicity.[28]

Rifapentine drug is a derivative of cyclo-pentylrifampicin. It binds to the β-subunit of the MTB RNA polymerase.[46] Combination of rifapentine and isoniazid for three months showed effective results against latent TB but ineffective against active TB.[71] Rifapentine is currently in the phase III clinical trial.

Although drug finding research is a highly convoluted, tedious and high-cost course, yet numerous new drugs for TB therapy have reached the preclinical and clinical trials with their mechanistic explanation. Nevertheless, novel drugs are required to battle the ever-growing menace of drug resistance.

9.4 PLANT-BASED NATURAL PRODUCTS IN RESEARCH ON TB DRUG

Since ancient times, various parts of plant and their extracts have been used as medicines against many diseases. This traditional knowledge may be beneficial in designing future potent drugs.[52] Plants have plentiful metabolites and similar biosynthetic pathways, which can be manipulated for health benefits. The explanation of genetic and biochemical means of natural products with curative action has proven to be priceless for natural and medicinal chemists as tools for developing novel front-line drugs.[48] Natural products have usually been the most frequent basis of drugs, and still contribute to 30% of the present pharmaceutical market.[34] Herbal products are key sources of novel therapeutic agents for bacterial, parasitic, fungal diseases, cancer, lipid disorders, and immunomodulation.[51]

Natural products continue to illuminate the development of new drugs to treat the TB patients. Numerous classes of anti-mycobacterial scaffolds from natural or herbal products have recently been reviewed.[61] Among all and flavonoids as anti-mycobacterial agents are gaining special attention for many reasons, such as (1) First, a huge variety of chemically diverse flavonoids is extensively present in standard human diet; (2) Second, as there are no consistent side effects, which have been linked with bioflavonoid consumption, high intake of dietary flavonoids is generally regarded as safe;

and (3) Third, flavonoids have been famous for their therapeutic implications and in boosting human health.[31] In the following sections, various types and sources of flavonoids have been discussed.

9.5 FLAVONOIDS

Flavonoids are natural phytochemicals that are present as polyphenolic secondary plant metabolites in different parts of the plant.[49] There are around 4000 varieties of flavonoids that are categorized into subclasses, such as flavonols, flavones, anthocyanins, isoflavones, and many more. The subclasses are further divided into various subtypes as illustrated in Table 9.1.

TABLE 9.1 Different Groups of Flavonoids.

Flavonoids	
Subclass	**Subtypes**
Anthocyanins	Cyanidin, delphinidin, pelargonidin, malvidin
Flavanols	Catechin, epicatechin, epigallocatechin, glausan-3-epicatechin, proanthocyanidins
Flavanones	Naringenin, hesperetin
Flavones	Luteolin, apigenin, tangeretin
Flavonols	Quercetin, kaempferol, myricetin
Isoflavones	Genistein, daidzein

Flavonoids are set of bioactive mixtures, which are widely present in natural products and the richest sources are: apples, green tea, red wine, onions, cranberries, and many more. Some flavonoids (such as flavanols, flavonols, and anthocyanins) are comparatively rich in human diet. The richest food sources of flavonoids based on their subclasses are categorized in Table 9.2.

In plants, flavonoids have numerous significant and vital roles, such as they provide defense guard toward damaging UV rays/plant pigmentation. Some of them also have the abilities to control gene expression and to modulate enzymatic action.[54] Moreover, several reports have shown that they possess antibacterial, free radical scavenging capacity, anticancer, antioxidant, coronary heart diseases protective, and antiviral properties. However, only few of them have been examined in depth.

TABLE 9.2   Different Subclasses of Flavonoids and Their Sources.

Subclass of flavonoids	Sources
Anthocyanins	Hazel nuts, pears, black grapes, strawberries, pecan nuts, red bilberries, red currants, aronia, raspberries, chickpeas
Flavanols	Apple juice, apricots, peaches, red wine, pecan nuts, broad beans, dark chocolate, black tea
Flavanones	Oranges, limes and lemons, grapefruit, dried oregano, artichokes
Flavones	Red grapes, chicory, lemons, kohlrabi, fresh parsley, green pepper
Flavonols	Apples, eggs, buckwheat, brussel sprouts, cranberries, asparagus, morello cherries, red onions
Isoflavones	Cocoa, fresh capers, sorrel, radish, green tea, brewed

9.6   FLAVONOIDS AS POTENTIAL TREATMENT FOR TB

9.6.1   CATECHIN

Catechin is a natural phenol and antioxidant, which belongs to flavonols class. It is a plant secondary metabolite that is derived from leaves of green tea (*Camellia sinensis*). There are mainly four types of catechins, such as epicatechin, epicatechin-3-gallate, epigallocatechin, and epigallocatechin-3-gallate. It positively affects several organ systems in our body. It exhibits numerous health benefits, such as anticarcinogenic, antidiabetic, anti-obesity, anti-artherogenic, hepatoprotection, neuroprotection, cardiovascular protection, and so forth. Catechins have direct antibacterial influence against broad spectrum of bacteria (e.g., *E. coli*, *Salmonella* spp., *S. aureus*, *Enterococcus* spp.). It is proved to hinder the bacterial binding to oral surfaces.[58]

Catechins bind to the lipid bilayer cell membrane of bacteria, which ultimately leads to bacterial membrane damage.[69] Injury to the bacterial membrane suppresses the capability of the microbe to attach to the host cells. It also suppresses the capability of the bacteria to adhere with each other therefore inhibiting the biofilm formation, which is the most important criterion in the pathogenicity of bacteria.[65] Thus, there is a high possibility that catechins might affect cell membrane of mycobacteria also.

Catechin has two rings of benzene (A and B) and a heterocyclic ring (C) having hydroxyl group attached at C3 position. Positions C2 and C3 possess chiral centers. It has four diastereoisomers, out of which two are cis-isomers called epicatechin, and other two are trans-isomers called catechin.

Mycobacterial cell wall is basically composed of three layers, namely: mycolic acid, arabinogalactan, and peptidoglycan. Fatty acids are essential component of mycolic acid layer. Additionally, fatty acids also serve as major source of bacterial energy. Therefore, targeting biosynthesis of fatty acids holds tremendous potential for anti-mycobacterial drug development.[80] Reports have shown that bioconstituents in green tea prohibit specific reductases, FabG and FabI, which are involved in synthesis of bacterial type II fatty acids.[84] It also hinders production of toxic metabolites by bacteria.[60]

Researchers have observed that crude green tea catechin extract exhibits a beneficial role in pulmonary tuberculosis (PTB) patients as adjuvant therapy in controlling the oxidative stress. In that study 2010, newly diagnosed AFB positive-PTB patients were selected and 500 µg of catechin with standard anti-tuberculosis treatment drugs was administered to them. After the first and fourth month of treatment, levels of oxidative stress were assessed and compared with the control (patients administered with standard anti-tuberculosis treatment drugs without catechin). The results revealed a significant decrease in levels of oxidative stress in cases when compared to control patients. Therefore, for reducing the oxidative stress in patients of PTB, crude green tea catechin can be administered as supportive therapeutic.[1]

9.6.2 EPIGALLO CATECHIN GALLATE (EGCG)

Epigallocatechin gallate is a phenolic antioxidant present in green and black tea. It is a chief form of catechin present in tea from *Camellia sinensis*. A content of 100–200 mg of EGCG may be present in a single cup of tea. It mainly acts as an antioxidant by inhibiting cellular oxidation and thereby prevents human cells from free radical damage. It has numerous health benefits, including antimicrobial activity against many bacteria and viruses. A decreased risk of TB disease is associated with high intake of tea.[11] Although EGCG is the foremost biologically potent catechin in tea, yet its stability as well as bioavailability is limited.

EGCG has a pyrogallol-type structure. It endures autooxidation causing formation of reactive oxygen species, leading to polymerization and decomposition.[59] EGCG prohibits the activity of tubercle bacillus-InhA, an enoyl-acyl carrier protein reductase, with an IC_{50} of 17.4 µM. EGCG abrogates the association of NADH with InhA. Fluorescence titration direct binding assays and biochemical assays by means of molecular docking also proves its inhibition of InhA.[66]

EGCG suppresses the persistence of mycobacterium tubercle bacilli inside macrophages by inhibiting the expression of TACO gene, which encodes for tryptophan-aspartate containing coat protein. Product of this gene supports both the entry and the intracellular survival of MTB within macrophages. Mycobacterial invasion experiments in THP-1 macrophages showed that EGCG has an intrinsic property of restricting the survival and entry of MTB into the host cells.[6] Therefore, in general, EGCG inhibits MTB survival within human macrophages.

Various reports have shown anti-mycobacterial activity. But interestingly, a study reported that EGCG gets degraded within a day if kept in the medium containing MTB bacteria. However, the degradant molecules of EGCG are actually comprised of anti-mycobacterial property. Therefore, if green tea contains degradant molecules of EGCG, it could be considered as a potential prophylactic agent against the MTB bacteria.[72]

A study demonstrated the effect of EGCG on the cell envelope structure of *Mycobacterial smegmatis*. By utilizing transmission electron microscopy, alteration in mycobacterial cell wall structure can be visualized after treatment with EGCG. The cell wall displayed a less electron-translucent region that became coarser and denser. Therefore, the study confirmed that EGCG affects the mycobacterial cell wall integrity and strengthens the probability of using EGCG as a prophylactic compound against tuberculosis.[72]

9.6.3 QUERCETIN

Quercetin is a flavonol and is extensively found in many fruits and vegetables, but apples and onions contain maximum level. Quercetin is a polyphenolic compound having many activities, namely, antimicrobial, anti-inflammatory, anticarcinogenic, and so forth. It has a mast cell stabilizing activity as well as it possess a cytoprotective effect on gastrointestinal cells.[53] Glycosylated forms of quercetin include rutin and quercetrin. These glycosylated forms have increased solubility, absorption as well as in vivo effects.[39]

Quercetin is comprised of the flavonol backbone (three carbon of the central ring is hydroxylated) along with two other hydroxylations on the outer ring. Removal of one of these hydroxyl groups generates kaempferol, which is the backbone for the icariin (active metabolite of Horny goat weed). Another metabolite, isorhamnetin is generated if the removed group gets substituted with a methoxy group.

There are various potential drug targets against MTB, which may be exploited for development of new compounds. One such target is DNA gyrase, as it is only present in prokaryotes. Quercetin blocks the activity of bacterial DNA gyrase, a type II DNA topoisomerase, in both *Mycobacterium smegmatis* and *Mycobacterium* tuberculosis. Docking studies demonstrated that quercetin interacts with Toprim domain of subunit B of DNA gyrase, which might be responsible for the action of quercetin. Using resazurin microtiter plate assay, a study estimated the MIC of quercetin as 100 µg/ml against the MTB. Therefore, report suggested that DNA gyrase could be a possible target for the quercetin in mycobacterium.[44,73]

A striking property of the Mycobacterium bacilli is that it can remain dormant during persistent infection. It can survive within the hostile environment of the macrophages in silent manner. An important enzyme of the glyoxylate shunt, Isocitrate lyase (ICL), is essential for survival of bacteria in the macrophages. Quercetin directly binds at the N-terminal region of ICL and thereby it inhibits the activity of ICL with IC_{50} of 3.57 µM.[68] Thus, quercetin is a promising drug against persistent tuberculosis infection.

First-line drugs against tuberculosis, isoniazid, and rifampicin, are known to cause severe hepatotoxicity. Noticeably, quercetin has been shown to impart hepatoprotective effects against the INH- and RFP-induced liver damage in experimental rats.[56] Therefore, quercetin has anti-mycobacterial effect and reduces hepatotoxicity. Owing to this dual effect, quercetin holds the potential of an ideal synergistic drug in the treatment of tuberculosis.

9.6.4 NARINGENIN

Naringenin is a natural flavonoid that is mainly found in grapes, tomatoes, and oranges. It belongs to flavanone subclass and has anti-mycobacterial, anti-inflammatory and antioxidant activities. Several studies revealed the molecular mechanism related to beneficial activities of naringenin, however, the evidence is still lacking.[2]

It has a backbone structure of flavanone with three hydroxy groups. It is present in two forms—aglycol and glycosidic, which has additional disaccharide neohesperidose connected through a glycosidic linkage. The naringenin-7-glucoside form has relatively less bioavailability as compared to aglycol form. The aglycon form of naringenin is readily formed in humans from its precursor.[2]

The two most important first-line drugs of anti-tubercular therapy; namely, Isoniazid (INH) and Rifampicin (RIF) are known to cause severe hepatic injury. The combinatorial administration of these drugs produces metabolic and morphological alterations in hepatic tissue.[62,79] Naringenin has shown to protect against numerous kinds of liver injury. In a recent study, researchers have evaluated the effects of Naringenin on INH- and RIF-induced liver damage in mice models. Their results suggest that Naringenin pretreatment significantly undermines the INH- and RIF-induced increase of the hepatic cell index, and serum ALT and AST (markers of liver damage). Therefore, Naringenin has strong hepato-protective effect against tuberculosis drug-induced hepatic injury.[83]

Furthermore, few studies have shown the involvement of oxidative stress in INH- and RIF-induced liver cell injury.[10,13] The results revealed that GHS (hepatic glutathione) content and SOD (superoxide dismutase: a marker for defense system against ROS) activity in liver cells were significantly reduced after INH and RIF administration. Contrastingly, there was a drastic increase in the levels of Malondialdehyde (MDA), which is a lipoperoxide marker of oxidative stress and the antioxidant status. Naringenin possesses potent antioxidant activity. It reduces oxidative stress through lowering the lipid peroxide level in liver and kidney tissue of mice models.[10] These observations affirm that Naringenin can be used to reduce the oxidative stress in liver and kidney cells, concomitant with INH and RIF treatment against tuberculosis.

9.6.5   APIGENIN

Apigenin is a trihydroxyflavone compound that is present in chamomile tea. It exhibits anti-mycobacterial, antibacterial, anti-inflammatory and anticancer activities. Inflammation is a defensive process facilitated with several chemical factors, which help to eradicate the early onset of diseases.[33] Tuberculosis is carried through acute and/or chronic inflammatory mechanisms by large generation of chemical agents that promote anti-inflammatory process.[27] The inflammation drugs possess side effects related with cardiovascular and gastric systems. Therefore, researchers have used methanol extract of *Schinus terebinthifolius*, fraction A3, and apigenin to evaluate anti-inflammatory property, nitric oxide (NO) generation prohibitory as well as for antioxidant activity in vitro. Researchers found that at IC_{50} of 9.25 ± 1.24 µg/ml, their mixture showed the suppression of NO generation through Lipopolysaccharide-stimulated macrophages as well as free radical scavenging, which contributed to anti-inflammatory properties against MTB.[9]

9.6.6   OTHER BIOFLAVONOIDS AS ANTI-TUBERCULAR AGENTS

Various flavonoids (such as isobachalcone, kanzanol C, 4-hydroxyloncho-carpin, stipulin, and amentoflavone) isolated from *Dorstenia barteri* showed activity against MTBH37Rv and *Mycobacterium smegmatis*. The IC_{50} value was estimated to be 2.44 µg/mL to 30 µg/mL.[36]

Phytocompounds (viz. cinnamolyglicoflavonoids 3-cinnamoyltribuloside, afzein, and stilbin) isolated from ethanol leaf extracts of *Heritiera littoralis* showed anti-mycobacterial activity against *Myobacterium madagascarience* and *Myobacterium indicus pranii*. The IC_{50} value was estimated as 5.0 mg/mL and 1.6 mg/mL to 0.8mg/mL for the crude extracts and for the purified compounds, respectively.[14]

Flavonoids (such as mombinrin (a coumarin), mombincone, mombi-noate, and mombinol) derived from *Spondias mombin* also showed anti-mycobacterial activity against MTB at IC_{50} value of 40 µg/mL.[50]

Kenusanone-F-7-methyl ether and Sophoronol-7-methyl ether belong to 3-hydroxyisoflavanones subtypes, which are isolated from stem bark of *Dalbergia melanoxylon*. They showed anti-tubercular activity against MTB H37Rv. Kenusanone-F-7-methyl ether exhibited 96% inhibition at 128 µM having IC_{50} value of 80.3 µM. Docking studies also showed that both of these agents possess great affinity for the MTB InhA protein.[47]

Cinnamic acid has been used to treat TB since 19th century. The IC_{50} value of cinnamic acid was estimated as 270 µM.[29] In the development of drugs against TB, some of the substituted classes of cinnamic acids have been reported as novel scaffolds.[18] Cerulenin (CRL) and trans-cinnamic acid (TCN) are two substituted forms of cinnamic acid. The IC_{50} value of CRL and TCN were estimated to be 5 µg/mL and 100 µg/mL, respectively. Anti-TB drugs (such as rifampicin, ethambutol, amikacin, and clofazimine) showed synergistic effect with CRL and TCN against MTB H37Rv and strains of MDR-TB.[57]

Classes of flavonoids (viz. 5,4'-dihydroxy-3,7,8,3-tetramethoxyflavone and 5,4'-dihydroxy-3,7,8-trimethoxyflavone) from *Larrea tridentate* for MDR-TB showed MIC value of 25 µg/mL and 25 µg/mL to 50 µg/mL, respectively.[26]

Research has been ongoing on natural flavonoids having anti-tubercular activity.[7] The characterization of flavonoids as novel anti-TB agents offers a promising alternative against tuberculosis. Furthermore, the substituted flavonoids provide additional advantages as they are shown to be active against drug resistant TB.

9.7 THERAPEUTIC IMPLICATIONS OF BIOFLAVONOIDS IN DISEASES OTHER THAN TUBERCULOSIS

9.7.1 EFFECT ON NERVOUS SYSTEM

Prevention of age-associated neurodegenerative disorders (e.g., dementia, Parkinson's, and Alzheimer's diseases) with the use of flavonoids have been explored recently. Reports suggest that flavonoids and its constituents can offer preservation of neuronal and cognitive brain functions.[55] Furthermore, flavonoids have been shown to provide neuronal protection and improvement in their regeneration processes.[76] *Ginkgo biloba* plant extract has high content of flavonoids, which has been useful in the treatment of dementia and Alzheimer's disease.[8] Another flavonoid, Tangeretin, mainly found in citrus fruits, belongs to the flavone subclass has shown to provide protection against Parkinson's disease.[17]

9.7.2 EFFECT ON CARDIOVASCULAR SYSTEM

The antioxidant activities of flavonoids have been effective to prevent cardiovascular diseases.[41] Antioxidant properties of flavonoid compounds depend both on polyphenol content as well as on their type. For example, catechin and quercetin exhibit the highest antioxidant activities in vitro.[22] Though, their metabolism in humans is still not clearly deciphered. One of the study significantly demonstrated health benefits of dietary flavonoids to decrease the cardiovascular-related mortality.[43]

Obesity is often related with dyslipidaemia, insulin resistance, and type 2 diabetic conditions.[64] Supplementation of flavonoid diet in obese or normal body mass mice showed better lipid profile when compared to the control with non-flavonoid diet. It also resulted in reduced insulin resistance and decreased visceral adipose tissue mass. These observations validated the role of flavonoids in protecting the cardiovascular system.[64]

9.7.3   EFFECT ON CANCER PREVENTION

Various reports have highlighted the role of flavonoids in chemoprevention. Cyclins and cyclin-dependent kinases (CDKs) are two types of regulatory molecules involved in the cell cycle progression. Activation of CDKs in an uncontrolled manner can lead to cancer. Therefore, to check this process, massive research has been focused on substances that can prevent or control CDKs activation. Many different flavonoids exhibited such activity, such as genistein, quercetin, daidzein, luteolin, kaempferol, apigenin, and epigallocatechin.[24]

The fundamental mechanisms of flavonoids[35] are needed to be deciphered properly. By considering the existing knowledge, a flavonoid-rich diet should be encouraged. It should be kept in mind that toxicity effects of diet with an excessive amount of flavonoid are largely unknown. Therefore, such dietary supplements should be consumed with caution.

9.8   SUMMARY

Mycobacterium tuberculosis (MTB) causes both pulmonary and extra-pulmonary tuberculosis. Current TB treatment regimen of nine months includes four primary drugs, such as isoniazid, rifampicin, ethambutol, and pyrazinamide. Although this treatment has a high success rate, yet the efficacy of this regime is restricted by defiance issues due to an extended period of hospitalization. Another important issue is the severe side effects associated with the drugs, which leads to the discontinuation of treatment by the patient. For these reasons, there is an imperative need for the expansion of potent and safe drugs to control TB.

Coumarin derivatives, phenolic compounds, and flavonoids have shown to lower down the burden of TB. Accumulating evidences support the fact that plant-based flavonoids may be better substitutes for the management and treatment of TB. Many fruits and vegetables are rich in flavonoids. Various flavonoids (viz. catechin, quercetin, naringenin, and many more) have been documented to possess anti-mycobacterial properties. This chapter focuses on the recent advances in the development of anti-tubercular drug from flavonoids and their derivatives to fight against tuberculosis.

KEYWORDS

- **antibacterial**
- **anti-mycobacterial activity**
- **apigenin**
- **catechin**
- **flavonoids**
- **herbal**
- **tuberculosis**

REFERENCES

1. Agarwal, A.; Prasad, R.; Jain, A. Effect of Green Tea Extract (Catechins) in Reducing Oxidative Stress Seen in Patients of Pulmonary Tuberculosis on DOTS Cat I Regimen. *Phytomedicin* **2010**, *17* (1), 23–27.
2. Alam, M. A.; Subhan, N.; Rahman, M. M.; Uddin, S. J.; Reza, H. M. Effect of Citrus Flavonoids, Naringin and Naringenin, on Metabolic Syndrome and Their Mechanisms of Action. *Adv. Nutr.* **2014**, *5* (4), 404–417.
3. Alifano, P.; Palumbo, C.; Pasanisi, D.; Tala, A. Rifampicin—Resistance, rpoB Polymorphism and RNA Polymerase Genetic Engineering. *J. Biotechnol.* **2015**, *202*, 60–77.
4. Allen, P. N.; Noller, H. F. Mutations in Ribosomal Proteins S4 and S12 Influence the Higher Order Structure of 16-S Ribosomal RNA. *J. Mol. Biol.* **1989**, *208* (3), 457–468.
5. Almeida Da Silva, P. E.; Palomino, J. C. Molecular Basis and Mechanisms of Drug Resistance in Mycobacterium Tuberculosis: Classical and New Drugs. *J. Antimicrob. Chemother.* **2011**, *66* (7), 1417–1430.
6. Anand, P. K.; Kaul, D.; Sharma, M. Green Tea Polyphenol Inhibits Mycobacterium Tuberculosis Survival within Human Macrophages. *Int. J. Biochem. Cell B.* **2006**, *38* (4), 600–609.
7. Askun, T. The Significance of Flavonoids as Potential Anti-Tuberculosis Compounds. *Res. Rev.: J. Pharm. Toxicol. Stud.* **2015**, *3* (3), 1–12.
8. Bastianetto, S.; Zheng, W. H.; Quirion, R. The Ginkgo Biloba Extract (EGb 761) Protects and Rescues Hippocampal Cells against Nitric Oxide-Induced Toxicity: Involvement of Its Flavonoid Constituents and Protein Kinase C. *J. Neurochem.* **2000**, *74* (6), 2268–2277.
9. Bernardes, N. R.; Heggdorne-araújo, M.; Borges, I. F. J. C. Nitric Oxide Production, Inhibitory, Antioxidant and Antimycobacterial Activities of the Fruits Extract and Flavonoid Content of *Schinus terebinthifolius*. *Rev. Bras. Farmacogn.* **2014**, *24* (6), 644–650.
10. Bhadauria, S.; Mishra, R.; Kanchan, R.; Tripathi, C.; Srivastava, A.; Tiwari, A.; Sharma, S. Isoniazid-Induced Apoptosis in HepG2 Cells: Generation of Oxidative Stress and Bcl-2 Down-Regulation. *Toxicol. Mech. Methods* **2010**, *20* (5), 242–251.

11. Chen, M.; Deng, J.; Li, W.; Lin, D.; Su, C. Impact of Tea Drinking upon Tuberculosis-neglected Issue. *BMC Public Health* **2015,** *15,* 515.

12. Chopra, P.; Meena, L. S.; Singh, Y. New Drug Targets for Mycobacterium Tuberculosis. *Indian J. Med. Res.* **2003,** *117,* 1–9

13. Chowdhury, A.; Santra, A.; Bhattacharjee, K.; Ghatak, S.; Saha, D. R.; Dhali, G. K. Mitochondrial Oxidative Stress and Permeability Transition in Isoniazid and Rifampicin Induced Liver Injury in Mice. *J. Hepatol.* **2006,** *45* (1), 117–126.

14. Christopher, R.; Nyandoro, S. S.; Chacha, M.; de Koning, C. B. A New Cinnamoyl Glycoflavonoid, Antimycobacterial and Antioxidant Constituents from *Heritiera littoralis* Leaf extracts. *Nat. Prod. Res.* **2014,** *28* (6), 351–358.

15. Culliton, B. J. Drug-Resistant TB May Bring Epidemic. *Nature* **1992,** *356* (6369), 473.

16. Daniel, T. M. Rifampin—Major New Chemotherapeutic Agent for the Treatment of Tuberculosis. *N. Engl. J. Med.* **1969,** *280* (11), 615–616.

17. Datla, K. P.; Christidou, M.; Widmer, W. W.; Rooprai, H. K.; Dexter, D. T. Tissue Distribution and Neuroprotective Effects of Citrus Flavonoid Tangeretin in a Rat Model of Parkinson's Disease. *Neuroreport* **2001,** *12* (17), 3871–3875.

18. De, P.; Koumba Yoya, G.; Constant, P.; Bedos-Belval, F.; Duran, H. Design, Synthesis, and Biological Evaluation of New Cinnamic Derivatives as Antituberculosis Agents. *J. Med. Chem.* **2011,** *54* (5), 1449–1461.

19. Deoghare, S. Bedaquiline—New Drug Approved for Treatment of Multidrug-Resistant Tuberculosis. *Indian J. Pharmacol.* **2013,** *45* (5), 536–537.

20. Desjardins, C. A.; Cohen, K. A.; Munsamy, V.; Abeel, T.; Maharaj, K. Genomic and Functional Analyses of Mycobacterium Tuberculosis Strains Implicate Ald in D-Cycloserine Resistance. *Nat. Genet.* **2016,** *48* (5), 544–551.

21. Diacon, A. H.; Dawson, R.; Hanekom, M.; Narunsky, K.; Maritz, S. J.; Venter, A.; Donald, P. R.; van Niekerk, C. Early Bactericidal Activity and Pharmacokinetics of PA-824 in Smear-Positive Tuberculosis Patients. *Antimicrob. Agents Chemother.* **2010,** *54* (8), 3402–3407.

22. Duthie, G.; Morrice, P. Antioxidant Capacity of Flavonoids in Hepatic Microsomes Is Not Reflected by Antioxidant Effects In Vivo. *Oxid. Med. Cell. Longev.* **2012,** *2012,* 165127.

23. Dutta, N. K.; Karakousis, P. C. Latent Tuberculosis Infection: Myths, Models, and Molecular Mechanisms. *Microbiol. Mol. Biol. Rev.* **2014,** *78* (3), 343–371.

24. Egert, S.; Rimbach, G. Which Sources of Flavonoids: Complex Diets or Dietary Supplements? *Adv. Nutr.* **2011,** *2* (1), 8–14.

25. Esmail, H.; Barry, C. E.; Young, D. B.; Wilkinson, R. J. The Ongoing Challenge of Latent Tuberculosis. *Philos. Trans. R. Soc. Lond. B, Biol. Sci.* **2014,** *369* (1645), Article ID: 20130437.

26. Favela-Hernandez, J. M.; Garcia, A.; Garza-Gonzalez, E. Antibacterial and Antimyco-bacterial Lignans and Flavonoids from *Larrea tridentata. Phytother. Res.* **2012,** *26* (12), 1957–1960.

27. Gaestel, M.; Kotlyarov, A.; Kracht, M. Targeting Innate Immunity Protein Kinase Signaling in Inflammation. *Nat. Rev. Drug Discov.* **2009,** *8* (6), 480–499.

28. Gler, M. T.; Skripconoka, V.; Sanchez-Garavito, E.; Xiao, H. Delamanid for Multidrug-Resistant Pulmonary Tuberculosis. *N. Engl. J. Med.* **2012,** *366* (23), 2151–60.

29. Guzman, J. D.; Mortazavi, P. N.; Munshi, T. 2-Hydroxy-Substituted Cinnamic Acids and Acetanilides are Selective Growth Inhibitors of Mycobacterium Tuberculosis. *Medchemcomm* **2014,** *5* (1), 47–50.

30. Haagsma, A. C.; Podasca, I.; Koul, A.; Andries, K. Probing the Interaction of the Diarylquinoline TMC207 with Its Target Mycobacterial ATP Synthase. *PLoS One* **2011**, *6* (8), E-Article: 23575.

31. Harborne, J. B. *The Flavonoids: Advances in Research since 1986*. Chapman & Hall: London, 1994; p. 241.

32. Hoagland, D. T.; Liu, J.; Lee, R. B.; Lee, R. E. New Agents for the Treatment of Drug-resistant Mycobacterium Tuberculosis. *Adv. Drug Del. Rev.* **2016**, *102*, 55–72.

33. Iwasaki, A.; Medzhitov, R. Regulation of Adaptive Immunity by the Innate Immune System. *Science* **2010**, *327* (5963), 291–295.

34. Kirkpatrick, P. Antibacterial Drugs—Stitching Together Naturally. *Nat. Rev. Drug Discov* **2002**, *1* (10), 748–748.

35. Kozlowska, A.; Szostak-Wegierek, D. Flavonoids: Food Sources and Health Benefits. *Rocz Panstw Zakl Hig* **2014**, *65*(2), 79–85.

36. Kuete, V.; Ngameni, B.; Mbaveng, A. T.; Ngadjui, B.; Meyer, J. J.; Lal, N. Evaluation of Flavonoids from *Dorstenia barteri* for Their Antimycobacterial, Antigonorrheal and Anti-Reverse Transcriptase Activities. *Acta Tropica* **2010**, *116* (1), 100–104.

37. Kumar, S.; Pandey, A. K. Chemistry and Biological Activities of Flavonoids: An Overview. *Sci. World J.* **2013**, *2013*, Article ID: 162750.

38. Lehmann, J. Para-Amino Salicylic Acid in the Treatment of Tuberculosis. *Lancet* **1946**, *1* (6384), 15–20.

39. Li, Y.; Yao, J.; Han, C.; Yang, J. Quercetin, Inflammation and Immunity. *Nutrients* **2016**, *8* (3), 167–172.

40. Machado, D.; Perdigao, J.; Ramos, J.; Couto, I.; Portugal, I. High-Level Resistance to Isoniazid and Ethionamide in Multidrug-Resistant Mycobacterium Tuberculosis of the Lisboa Family is Associated with inhA Double Mutations. *J. Antimicrob. Chemother.* **2013**, *68* (8), 1728–1732.

41. Majewska-Wierzbicka, M.; Czeczot, H. Flavonoids in the Prevention and Treatment of Cardiovascular Diseases. *Polski merkuriusz lekarski: Organ Polskiego Towarzystwa Lekarskiego* **2012**, *32* (187), 50–54.

42. Manjunatha, U.; Boshoff, H. I.; Barry, C. E. The Mechanism of Action of PA-824: Novel Insights from Transcriptional Profiling. *Commun. Integr. Biol.* **2009**, *2* (3), 215–218.

43. McCullough, M. L.; Peterson, J. J.; Patel, R.; Jacques, P. F. Flavonoid Intake and Cardiovascular Disease Mortality in a Prospective Cohort of US Adults. *Am. J. Clin. Nutr.* **2012**, *95* (2), 454–464.

44. Mdluli, K.; Ma, Z. Mycobacterium Tuberculosis DNA Gyrase as a Target for Drug Discovery. *Infect. Disord. Drug Targets* **2007**, *7* (2), 159–168.

45. Meintjes, G. Management of Drug-Resistant TB in Patients with HIV Co-Infection. *J. Int. AIDS Soc.* **2014**, *17* (4 Supplement 3), Article ID: 19508.

46. Munsiff, S. S.; Kambili, C.; Ahuja, S. D. Rifapentine for the Treatment of Pulmonary Tuberculosis. *Clin. Infect. Dis. Off. Pub. Infect. Dis. Soc. Am.* **2006**, *43* (11), 1468–1475.

47. Mutai, P.; Heydenreich, M.; Thoithi, G.; Mugumbate, G. 3-Hydroxyisoflavanones from the Stem Bark of *Dalbergia melanoxylon*: Isolation, Antimycobacterial Evaluation and Molecular Docking Studies. *Phytochem. Lett.* **2013**, *6*, 671–675.

48. Newman, D. J.; Cragg, G. M.; Snader, K. M. The Influence of Natural Products upon Drug Discovery. *Nat. Prod. Rep.* **2000**, *17* (3), 215–234.

49. Nijveldt, R. J.; van Nood, E.; van Hoorn, D. E.; Boelens, P. G.; van Norren, K.; van Leeuwen, P. A. Flavonoids: Review of Probable Mechanisms of Action and Potential Applications. *Am. J. Clin. Nutr.* **2001**, *74* (4), 418–425.

50. Olugbuyiro, J.O.; Moody, J.O. Anti-Tubercular Compounds from *Spondias mombin*. *Int. J. Pure App. Sci. Technol.* **2013,** *19* (2), 76–87.

51. Pan, S. Y.; Zhou, S. F.; Gao, S. H.; Yu, Z. L. New Perspectives on How to Discover Drugs from Herbal Medicines: CAM's Outstanding Contribution to Modern Therapeutics. *Evidence-based Complementary and Alternative Medicine:* eCAM **2013,** *2013,* Article ID: 627375.

52. Pauli, G. F.; Case, R. J.; Inui, T.; Wang, Y.; Cho, S.; Fischer, N. H.; Franzblau, S. G. New Perspectives on Natural Products in TB Drug Research. *Life Sci.* **2005,** *78* (5), 485–494.

53. Penissi, A. B.; Rudolph, M. I.; Piezzi, R. S. Role of Mast Cells in Gastrointestinal Mucosal Defense. *Biocell: Official J. Sociedades Latinoamericanas de Microscopía Electronica* **2003,** *27* (2), 163–172.

54. Pollastri, S.; Tattini, M. Flavonols: Old Compounds for Old Roles. *Ann. Bot.* **2011,** *108* (7), 1225–1233.

55. Prasain, J. K.; Carlson, S. H.; Wyss, J. M. Flavonoids and Age-related Disease: Risk, Benefits and Critical Windows. *Maturitas* **2010,** *66* (2), 163–171.

56. Qader, G. I.; Aziz, R. S.; Ahmed, Z. A.; Abdullah, Z.; Hussain S. A. Protective Effects of Quercetin against Isoniazid and Rifampicin Induced Hepatotoxicity in Rats. *Am. J. Pharmacol. Sci.* **2014,** *2* (3), 56–60.

57. Rastogi, N.; Goh, K. S.; Horgen, L.; Barrow, W. W. Synergistic Activities of Antituberculous Drugs with Cerulenin and Trans-Cinnamic Acid against Mycobacterium Tuberculosis. *FEMS Immunol. Med. Mic.* **1998,** *21* (2), 149–157.

58. Reygaert, W. C. The Antimicrobial Possibilities of Green Tea. *Front. Microbiol.* **2014,** *5,* 434.

59. Saeki, K.; Hayakawa, S.; Isemura, M.; Miyase, T. Importance of a Pyrogallol-Type Structure in Catechin Compounds for Apoptosis-Inducing Activity. *Phytochemistry,* **2000,** *53* (3), 391–394.

60. Sakanaka, S.; Okada, Y. Inhibitory Effects of Green Tea Polyphenols on the Production of a Virulence Factor of the Periodontal-Disease-Causing Anaerobic Bacterium *Porphyromonas gingivalis. J. Agr. Food Chem.* **2004,** *52* (6), 1688–1692.

61. Salomon, C. E.; Schmidt, L. E. Natural Products as Leads for Tuberculosis Drug Development. *Curr. Top. Med. Chem.* **2012,** *12* (7), 735–765.

62. Santosh, S.; Sini, T. K.; Anandan, R.; Mathew, P. T. Hepatoprotective Activity of Chitosan against Isoniazid and Rifampicin-Induced Toxicity in Experimental Rats. *Eur. J. Pharmacol.* **2007,** *572* (1), 69–73.

63. Selikoff, I. J.; Robitzek, E. H.; Ornstein, G. G. Treatment of Pulmonary Tuberculosis with Hydrazide Derivatives of Isonicotinic Acid. *JAMA* **1952,** *150* (10), 973–980.

64. Shabrova, E. V.; Tarnopolsky, O.; Singh, A. P.; Plutzky, J.; Vorsa, N.; Quadro, L. Insights into the Molecular Mechanisms of the Anti-Atherogenic Actions of Flavonoids in Normal and Obese Mice. *PLoS One* **2011,** *6* (10), E-article: 24634.

65. Sharma, A.; Gupta, S.; Sarethy, I. P.; Dang, S.; Gabrani, R. Green Tea Extract: Possible Mechanism and Antibacterial Activity on Skin Pathogens. *Food Chem.* **2012** *135* (2), 672–675.

66. Sharma, S. K.; Kumar, G.; Kapoor, M.; Surolia, A. Combined Effect ofEpigallocatechin Gallate and Triclosan on Enoyl-ACP Reductase of Mycobacterium Tuberculosis. *Biochem. Bioph. Res. Co.* **2008,** *368* (1), 12–17.

67. Shi, W.; Zhang, X.; Jiang, X.; Yuan, H. Pyrazinamide Inhibits Trans-Translation in Mycobacterium Tuberculosis. *Science* **2011,** *333* (6049), 1630–1632.

68. Shukla, H.; Kumar, V.; Singh, A. K.; Rastogi, S. Isocitrate Lyase of Mycobacterium Tuberculosis Is Inhibited by Quercetin through Binding at N-terminus. *Int. J. Biol. Macromol.* **2015,** *78,* 137–141.

69. Sirk, T. W.; Brown, E. F.; Friedman, M.; Sum, A. K. Molecular Binding of Catechins to Biomembranes: Relationship to Biological Activity. *J. Agr. Food Chem.* **2009,** *57* (15), 6720–6728.

70. Sreevatsan, S.; Pan, X.; Stockbauer, K. E.; Williams, D. L. Characterization of rpsL and rrs Mutations in Streptomycin-Resistant Mycobacterium Tuberculosis Isolates from Diverse Geographic Localities. *Antimicrob. Agents Chemother.* **1996,** *40* (4), 1024–1026.

71. Sterling, T. R.; Villarino, M. E.; Borisov, A. S.; Shang, N.; Gordin, F.; Bliven-Sizemore, E.; Hackman, J. Three Months of Rifapentine and Isoniazid for Latent Tuberculosis Infection. *N. Engl. J. Med.* **2011,** *365* (23), 2155–2166.

72. Sun, T.; Qin, B.; Gao, M.; Yin, Y. Effects of Epigallocatechin Gallate on the Cell-Wall Structure of Mycobacterial *smegmatis* mc(2)155. *Nat. Prod. Res.* **2015,** *29* (22), 2122–2124.

73. Suriyanarayanan, B.; Shanmugam, K.; Santhosh, R. S. Synthetic Quercetin Inhibits Mycobacterial Growth Possibly by Interacting with DNA Gyrase. *Rom. Biotech. Lett.* **2013,** *18* (5), 8587–8593.

74. Telenti, A.; Philipp, W. J.; Sreevatsan, S. The emb operon—A Gene Cluster of Mycobacterium Tuberculosis Involved in Resistance to Ethambutol. *Nat. Med.* **1997,** *3* (5), 567–570.

75. Udwadia, Z. F.; Amale, R. A. Totally Drug-Resistant Tuberculosis in India. *Clin. Infect. Dis.* **2012,** *54*(4), 579–581.

76. Vauzour, D.; Vafeiadou, K.; Rodriguez-Mateos, A. The Neuroprotective Potential of Flavonoids: A Multiplicity of effects. *Genes Nutr.* **2008,** *3* (3–4), 115–126.

77. Velayati, A. A.; Masjedi, M. R.; Farnia, P.; Tabarsi, P. Emergence of New Forms of Totally Drug-Resistant Tuberculosis Bacilli: Super Extensively Drug-Resistant Tuberculosis or Totally Drug-Resistant Strains in Iran. *Chest* **2009,** *136* (2), 420–425.

78. Von Groll, A.; Martin, A.; Jureen, P.; Hoffner, S. Fluoroquinolone Resistance in Mycobacterium Tuberculosis and Mutations in gyrA and gyrB. *Antimicrob. Agents Chemother.* **2009,** *53* (10), 4498–4500.

79. Wang, C.; Fan, R. Q.; Zhang, Y. X.; Nie, H.; Li, K. Naringenin Protects against Isoniazid-and Rifampicin-Induced Apoptosis in Hepatic Injury. *World J. Gastroenterol.* **2016,** *22* (44), 9775–9783.

80. Wang, Y.; Ma, S. Recent Advances in Inhibitors of Bacterial Fatty Acid Synthesis Type II (FASII) System Enzymes as Potential Antibacterial Agents. *Chem. Med. Chem.* **2013,** *8* (10), 1589–1608.

81. Wolucka, B. A. Biosynthesis of D-arabinose in mycobacteria: Novel Bacterial Pathway with Implications for Antimycobacterial Therapy. *FEBS J.* **2008,** *275* (11), 2691–2711.

82. World Health Organization (WHO). *Global Tuberculosis Report 2016.* World Health Organization: Geneva, 2016; p. 68.

83. Yen, H. R.; Liu, C. J.; Yeh, C. C. Naringenin Suppresses TPA-Induced Tumor Invasion by Suppressing Multiple Signal Transduction Pathways in Human Hepatocellular Carcinoma Cells. *Chem.-Biol. Interact.* **2015,** *235,* 1–9.

84. Zhang, Y. M.; Rock, C. O. Evaluation of Epigallocatechin Gallate and Related Plant Polyphenols as Inhibitors of the FabG and FabI Reductases of Bacterial Type II Fatty Acid Synthase. *J. Biol. Chem.* **2004,** *279* (30), 30994–31001.

85. Zhang, Y.; Shi, W.; Zhang, W.; Mitchison, D. Mechanisms of Pyrazinamide Action and Resistance. *Microbiol. Spectr.* **2014,** 2 (4), 1–12. Online: doi: 10.1128/microbiolspec. MGM2-0023-2013

86. Zhao, X.; Yu, H.; Yu, S.; Wang, F. Hydrogen Peroxide-Mediated Isoniazid Activation Catalyzed by Mycobacterium Tuberculosis Catalase-Peroxidase (KatG) and Its S315T Mutant. *Biochemistry* **2006,** *45* (13), 4131–4140.

87. Zheng, J.; Rubin, E. J.; Bifani, P.; Mathys, V. Para-Amino Salicylic Acid is a Prodrug Targeting Di-hydrofolate Reductase in Mycobacterium Tuberculosis. *J. Biol. Chem.* **2013,** *288* (32), 23447–23456.

88. Zumla, A.; Nahid, P.; Cole, S. T. Advances in the Development of New Tuberculosis Drugs and Treatment Regimens. *Nat. Rev. Drug Discov.* **2013,** *12* (5), 388–404.

CHAPTER 10

TRADITIONAL MEDICINAL PLANTS FOR RESPIRATORY DISEASES: MEXICO

ARMANDO ENRIQUE GONZÁLEZ-STUART and JOSÉ O. RIVERA

ABSTRACT

Medicinal plants have been used by all cultures throughout history and continue to be an integral part of the development of modern civilization. Mexicans and other Hispanic groups have used plants to treat various respiratory problems from generation to generation. The main afflictions, for which various plants are ingested (either taken as tea made from a single plant—or multiple combinations), include asthma, bronchitis, coughs and cold, pneumonia, and tuberculosis, etc. The meteoric surge of complementary and alternative medicine (CAM, currently known as integrative medicine), and the dissemination of the Hispanic population and culture throughout North America and Mexico, have introduced various species of medicinal herbs that may be scarcely known to Western medical practice. Therefore, both physicians and patients must be aware of potential risks and health benefits of using herbal medicines.

10.1 INTRODUCTION

Large portion of the population in neighboring cities El Paso, USA and Ciudad Juarez, Mexico, (which together comprise the largest international border in the world), commonly uses diverse medicinal plants.[1,2] Several types of herbs and herbal products in diverse forms (such as infusions (teas), extracts, capsules, tinctures, and, more recently, essential oils) have been used to treat various respiratory ailments, including asthma. Relatively few studies have assessed the importance, identity, and mode of use of medicinal plants from Mexico for the treatment of respiratory ailments along the international border between the United States and Mexico. Herbal products among the general US population can vary from slightly more than 10% to more than 50% within the international border setting. Research has shown that the use of medicinal plants tends to be higher among the Mexican Americans along this border, compared to the rest of the US. This includes herbs used for the treatment of various respiratory ailments and asthma.[1,2] Approximately, 3000–5000 medicinal plants are commonly used by traditional healers of Mexico to treat various diseases. Although various medicinal plants (such as eucalyptus, fennel, licorice, cinnamon, cassia, colt's foot, marjoram, and oregano) have been imported from other continents, and many native species are still being used by indigenous healers to treat diverse infectious or allergic afflictions related to the respiratory tract.[3,4]

This chapter reviews most commonly used medicinal plants employed by Mexican Traditional Medicine (MTM), with special emphasis on certain indigenous species, to treat various diseases associated with respiratory tract. The possible health benefits or potential health risks are also discussed. The in-depth treatment of specific diseases, for example, tuberculosis, is beyond the scope of this chapter and therefore is only mentioned briefly.

10.2 STATUS OF RESPIRATORY DISEASES: THE UNITED STATES AND MEXICO

Of the 15 leading causes of death in the United States in 2014, chronic lower respiratory diseases (CLRD) were third on the list and influenza and pneumonia were eighth. Various manifestations of the wide array of respiratory diseases can affect people at the individual, as well as the family level. The diverse forms of respiratory ailments can also take place in various environments, including, workplaces, churches, schools, communities, cities, and states. According to data from 2013, CLRD comprised the third leading cause of death in the US in 2010. Additionally, in a 2-year period (2007–2009), 11.8 million adults were diagnosed with chronic obstructive pulmonary disease (COPD), which is the primary factor associated with CLRD mortality. Previous research showed that approximately equal numbers to those diagnosed with COPD had not yet been diagnosed.[5] In 2010, the Pan American Health Organization (PAHO) estimated that certain respiratory afflictions (including influenza and pneumonia, other related complications) affected the very young, as well as the old in Mexico.[6]

10.3 MEDICINAL PLANT BIODIVERSITY

Mexico is home to a rich diversity of plants, with approximately 23,000 species,[7,8] which corresponds to about 10% of the world's flora. More than half of the world's population (65%) uses some sort of traditional medicine as a form of primary care to treat various health conditions.[9] In some countries, approximately half of the pharmaceutical products on the market today come from medicinal plant sources.[10] Of the 252 medications considered by the World Health Organization (WHO) to be essential, are derived from flowering plants.[11] Well into the 21st Century, millions of people around the globe continue to use medicinal plants as an alternative for the treatment of disease, sometimes due to the lack of modern medicines.[12]

10.4 INDIGENOUS MEXICAN HERBAL TRADITION

MTM is a combination of various healing modalities, including African, Asian, European, and native indigenous medicine.[13,14] Various remedies used by the ancient indigenous people of Mexico are still used on both sides of the US–Mexico border to treat a wide array of ailments, including respiratory diseases.[15,16]

10.4.1 USE OF HERBAL MEDICINE IN MEXICAN-AMERICAN CULTURE

MTM includes the use of diverse healing products comprising of a wide array of plants, fungi, animals, and minerals. Other centuries old healing traditions from Ayurveda and Traditional Chinese Medicine (TCM) also include various medicinal products derived from plants, animals, fungi, and minerals. Mexicans and Mexican Americans currently residing along the extensive US–Mexico border commonly use certain plants for the treatment of various respiratory problems, including asthma, coughs, colds, pneumonia, and tuberculosis. Infusions (herbal teas) as well as ointments and other diverse plant-based remedies, comprise an important character-istic of the international border culture.[17-20]

10.4.2   RESEARCH STUDIES: EFFICACY AND SAFETY OF HERBAL PLANTS IN MEXICO

A great number of medicinal plants from Mexico have not been studied in depth. Of the 3000–5000 plants used in MTM, only limited number have been studied regarding their active constituents, efficacy, and safety. In addi-tion, various introduced plants from Asia, Africa, and Europe, for example, have replaced many of the indigenous plants used by the native peoples of Mexico before the European conquest.[3]

The lack of research on potential medicinal properties of the plant kingdom is not only akin to Mexico, it is also a global problem. For example, of the approximately 250,000 known flowering plant species, only approximately 10% have been adequately studied.[21] Additionally, many people espouse the notion that because plants are "natural" they are inherently safe and free from any serious side effects, in sharp contrast to many over the counter (OTC) and prescription medications. Even though the exact application of certain species may have changed over the centuries, various medicinal plants are still used today along the US–Mexico border much in the same way they were employed in ancient Mesoamerica.[20,22,23]

10.4.3   ORIGIN OF MEDICINAL PLANTS: THE US–MEXICO BORDER

The use of herbs and related supplements in the US is from 13 to 19% within the general population.[24,25] In comparison, studies done in the largest international border of El Paso, Texas, and Ciudad Juarez, Chihuahua, showed that up to 59% of the participants mentioned using various herbal products within the previous 12 months, the majority of whom did not inform their healthcare providers about use of herbs or any other CAM therapy. Despite the limited number of studies available regarding the use of medicinal plants among the US–Mexico border population, the results mention that there is a higher prevalence of herbal products among the people of Hispanic descent.[1,26]

10.4.4 AVAILABILITY AND COST OF HERBAL PRODUCTS: US–MEXICO BORDER

Herbal products commercialized in the form of crude drugs (unprocessed parts of leaves, flowers, or roots, for example) or teas, are usually far less expensive than many prescription medications. They are also readily available on both sides of the international border without the need for a medical prescription. This is due in part to the October of 1994 act of the United States Congress known as the Dietary Supplement and Health Education Act (DSHEA). This act was in response to the humongous surge in popularity of various herbal remedies throughout the country. The DSHEA classified herbal and nutritional supplements as foods, not drugs, thus putting them out of reach of scrutiny of the Food and Drug Administration (FDA). This permitted the active commercialization of a plethora of herbal supplements in various forms, making it easier for these commodities to be integrated into the mainstream market, and available to the general public as OTC products, without the need for a medical prescription.[27,28]

10.5 SELECTED MEDICINAL PLANTS IN MTM

The following plants are used available in various parts of Mexico to treat diverse diseases related to the respiratory tract. Some are indigenous and have been used presumably for many centuries, while others were introduced into Mexico during the colonial period, when it was known as New Spain.[3,4]

10.5.1 MORMON TEA (EPHEDRA NEVADENSIS)

Also known by various common names in Spanish, such as Canutillo, Popotillo, and Tepopote, Mormon tea has been used in Northern Mexico and the Southwestern United States for many years to treat asthma and rhinitis. This North American species of Ephedra is supposed not to contain any of the proto-alkaloids, such as ephedrine and pseudoephedrine of their Asian counterparts.[29] However, one researcher contends that this species does indeed act as a nasal decongestant, and perhaps by certain phytochemicals are yet to be discovered.[30] The plant stalks are usually boiled in water to make a tea. Additionally, alcohol-based tinctures are also used in traditional medicine. According to certain sources, a related North American species (*Ephedra trifurca*) does contain ephedrine as its principal active ingredient.[31]

10.5.2   HOREHOUND (MARRUBIUM VULGARE)

A tea made from the leaves of this European plant is taken to treat sore throat and cough.[32] In Mexican popular medicine, the herb is valued for its diuretic, anti-inflammatory, expectorant, and fever-lowering properties. However, its use is contraindicated during pregnancy.[10,12]

10.5.3   OREGANO (ORIGANUM VULGARE)

Although not native to Mexico, oregano has been used for centuries in MTM to treat acute cough, as an anti-inflammatory for bronchitis, as well as against lung inflammation. The remedy consists of slowly heating the oregano leaves in water, but being careful not to bring it to a boil, since the resulting decoction could be harmful to the patient's health. In cases of severe cough, honey, and key-lime juice can be added to improve the tea's palatability and efficacy.[12]

10.5.4   LUNGWORT LICHEN (LOBARIA PULMONARIA)

This plant, known in English as "lungwort lichen," is a mutualistic symbiotic association between an algae and a fungus. It should not be confused with another plant of European origin (*Pulmonaria officinalis*), also commonly known as "lungwort," belonging to a different botanical family (Boragina-ceae), which is also used for respiratory problems.[33] In MTM, the whole plant is considered to possess anti-inflammatory actions, and is used as a tea for the treatment of cough, asthma, bronchitis, and pneumonia. Approxi-mately, 10–15 g of the plant is decocted in water for 5 or 6 min. The resulting decoction is strained and taken 4 times a day for a period of 15–20 days. The active ingredients include various phytosterols. An acetone extract of the plant can inactivate prions (self-replicating infective proteins) when applied to mice and hamsters.[34]

10.5.5   ICELAND MOSS (CETRARIA ISLANDICA)

The traditional use of this plant in Mexico as well as other countries is for the treatment of bronchitis cough, laryngitis, pharyngitis, and tuberculosis. The whole plant is used to make a tea. The plant is rich in mucilage and various other phytochemicals including usnic acid. The mucilaginous components act as antitussives, diuretics, and demulcents. The latter effects are useful to protect the respiratory mucosa. The lichenic acids have an expectorant as well as antibacterial and antifungal action. Herbalists in Mexico recommend decocting the 10–15 g of the plant in 3 L (approximately 3 quarts) of water during 2–4 min at a low heat. The decoction is strained and taken 4 times per day for a period of 15–20 days. Ingested in high doses, the plant could be irritating to the stomach.[34]

10.5.6 MEXICAN HAWTHORN (CRATAEGUS SPP)

This small spiny tree has been used to treat cough and cold since pre-Hispanic times. The most commonly used part of the plant is the fruit, although the roots, leaves, and flowers are also used. Small section of the root is cut into small pieces and boiled in water. The resulting tisane or tea is ingested for the treatment of cough and bronchitis. Approximately, 10–15 g of the plant is decocted in 3 L of water. The tea can be taken up to 4 times a day for a period of 2–3 weeks.[34] Another option is to decoct Mexican hawthorn fruits along with Poinsettia leaves and Mexican elder flowers (*Sambucus mexicana*), in half a liter of water. The resulting tea is sweetened with sugar and can be taken twice a day; one cup in the morning and one cup at night, for a period of 3–4 days.[4] The combination with elder flowers makes sense, as various species of the genus *Sambucus* have been shown to possess important antiviral activities.[35] Cinnamon and honey may also be added to a fruit decoction, to improve its taste and augment its efficacy. In order to treat the common cold, a decoction of three or four fruits is made (the flowers and leaves can also be used), along with three leaves of the Cherimoya (*Anona cherimoya*) tree. The tea can be taken hot or cold several times during the day.[36]

10.5.7 FAT WOLF (PSEUDOGNAPHALIUM OBTUSIFOLIUM [SYN. GNAPHALIUM OBTUSIFOLIUM])

Everlasting can be found in various regions of the US southwest and Northern Mexico.[37] In Mexico, the common name of "gordolobo" ("fat wolf") is given to many plants included in the *Senecio*, *Pseudognaphalium*, and *Gnaphalium* genera. All of these belong to the Asteraceae or Compositae botanical family.[38] The original medicinal plant brought by the Europeans, commonly known in Spanish as "gordolobo" is a very different species (*Verbascum thaspsus*-Scrophulariaceae), known in English as "mullein." Currently, the European species is still found in various parts of Mexico. However, the indigenous plants included within the Gnaphalium and *Pseudognaphalium* genera are much more commonly used by the Mexican population to treat various respiratory ailments.[3,39]

This species, also known in the United States as "cudweed," is one of the most sought-after herbs for the treatment of respiratory problems in Mexico. Its use for the treatment of various respiratory and digestive problems predates to the European conquest.[19] Gordolobo and related species are mainly employed for the treatment of respiratory problems, especially bronchitis and asthma.[40] Due to its antiviral and antiseptic properties, Gordolobo tea is also useful for the treatment of a variety of issues related to respiratory problems including emphysema, colds, sinusitis, and pneumonia, among others. The flowers and stems are used to make teas, which have anti-inflammatory actions, possibly reducing the effects of asthma. The tea is usually sweetened with honey. Laboratory tests have shown that various herbs within the *Gnaphalium* genus do indeed possess antimicrobial, anti-inflammatory, and antitussive actions, among others.[41]

The stems, leaves, and flowers are used and taken as an infusion (tea) to treat various respiratory problems, including asthma. The tea is sometimes spiced with cinnamon, which may give it a more potent effect. Milk and honey are sometimes added to the tea for the treatment of asthma, bronchitis, chest colds, cough, and sore and inflamed throat. Garlic and green tomato rind (*Physalis* spp) are sometimes added to the tea in order to treat bronchitis. The stems are decocted in water and taken along with warm milk as a tea for cough. However, if the cough is productive (i.e. with the presence of phlegm) then the tea should be made with water only. Gordolobo can be combined with various other herbs, such as Bougainville (*Bougainvillea* spp.), Krameria (*Krameria pauciflora*), Eucalyptus leaves (*Eucalyptus globulus*), Pine flowers (*Pinus* spp.), and Mexican hawthorn (*Crataegus mexicana*). The herbs are boiled in a liter of water and taken as a tea.[4,12]

For the treatment of cough with a nasal congestion, a decoction is made of six stems per liter of water and then taken a tea continuously during the day. When there is no nasal congestion, the tea is taken only twice a day. The tea is emollient and pectoral, since it can help relieve the symptoms of sore throat and chest pain associated with bronchitis.[4] For both dry as well as productive cough, the leaves and stems are decocted in water and taken along with honey and lime juice. As an expectorant, the leaves and stems are boiled in water and the patient inhales the vapors in order to expel the phlegm.[12]

Interestingly, similar species are also used elsewhere in Spanish America for the treatment of respiratory illnesses. In the traditional indigenous medicine of Chile, for example, teas are made from three related species, such as *G. glandulosum*, *G. lacteum*, and *G. vira vira*, which are used for the treatment of pulmonary and bronchial issues.[39,42] However, plants belonging to the aforementioned genus produce dangerous and carcinogenic hepatotoxins, known as pyrrolizidine alkaloids. These compounds are metabolized in the liver and pose a direct threat to humans as well as animals.[43,44] For this reason, it is not recommended to take this plant for prolonged treatments. Other herbs taken as teas (such as cinnamon, eucalyptus, and oregano) could be safer options for the long-term treatment of asthma and other bronchial problems.[45]

10.5.8 BOUGAINVILLEA (BOUGAINVILLEA SPECTABILIS)

This species is originally from Brazil, where it is valued as an attractive ornamental plant. However, Bougainvillea can be found in many tropical areas of the globe.[46,47] In Mexico, the flowers, alone or in combination with other plants, are decocted in water to make a tea for the treatment of cough and colds. Sometimes the stem bark may also be used. As an expectorant for productive cough, the vapor emanating from the decocted flowers is inhaled.[3,34,48] For bronchitis, the flowers are decocted in water along with a piece of ocote or "torch pine" (*Pinus teocote*), flowers of violet (*Viola* spp.), and a clove of garlic (*Allium sativum*). The decoction is taken as a tea especially at night, and can be sweetened with honey.[36] With regard to its efficacy and safety, there are no published studies to ascertain its efficacy for the treatment of respiratory problems, nor its safety if taken during gestation.

10.5.9 CALIFORNIA POPPY (ESCHSOLZIA CALIFORNICA)

Also known as "yellow poppy," this plant is native to Mexico (Northern Baja California) and is in the same plant family (Papaveraceae) as the opium poppy (*Papaver somniferum*), but does not contain that alkaloid. In traditional medicine, it is primarily used to treat cough due to its antitussive, sedative/analgesic, and antispasmodic properties. For the treatment of bronchitis and cough, a decoction of the flowers is taken as a tea until the symptoms are resolved. However, the tea should not be taken during pregnancy, as it may initiate labor.[49,50]

10.5.10 PORTER'S LOVAGE (LIGUSTICUM PORTERI)

The root and rhizomes of this plant are used medicinally. This species is also known in the United States as "oshá."[38] This plant has been used for centuries by diverse tribes in North America.[30,51] Chuchupate root has traditionally been taken as a tea for the treatment of a variety of ailments, including bronchitis, tuberculosis, pneumonia, colds, and sore throats. The root also possesses antiviral activity. It is also part of ritual curing ceremonies by various indigenous tribes of northern Mexico and the Southwestern United States.[52,53] The root contains various phytochemicals, notably monomeric, and dimeric phthalides including Z-ligustilide, Z-butylidenephthalide, tokinolide B, diligustilide, and riligustilide. Additionally, the plant contains phenolic compounds, such as coniferyl ferulate and ferulic acid, which confer antispasmodic and anti-inflammatory properties to the plant.[54] Chuchupate root should not be taken during pregnancy.

10.5.11 WILD BLACK CHERRY (PRUNUS SEROTINA)

The stems, bark, and leaves of this tree are used to treat cough. To make a cough syrup, two stems or six leaves are decocted in half a liter of water, adding sugar and cinnamon. The resulting liquid is given to children by one teaspoon until the symptoms resolve. Immature or tender leaves as well as the green fruits should not be consumed as they contain a compound known as "amygdalin," which is toxic due to the presence of hydrocyanic acid. Although the mature fruits can be safe to eat, their seeds may still contain hazardous levels of the toxic compound.[36]

10.5.12 JIMSON WEED (DATURA STRAMONIUM)

This and related species are known throughout the world by a variety of vernacular names.[38,39,55] This annual herb has been used as a medicine and in religious rituals since pre-Hispanic times by various indigenous tribes of Mexico (commonly known as "toloache") and other regions of Latin America. *D. stramomiun* has been used to treat various ailments, from muscular pain to asthma. This and other related specie of the genus *Datura* were included in America's first book on indigenous medicinal plants known as the *De la Cruz-Badiano manuscript* written in 1552.[56]

The genus *Datura* contains various species that are hallucinogenic and have been used for centuries in magic and healing rituals.[57] The medicinal and hallucinogenic uses of toloache were banned by the Mexican government many years ago, owing to the potential toxicity of its internal use. The plant contains various tropane alkaloids (including scopolamine and hyoscyamine), found in other species of the nightshade family, to which it belongs. For many decades, *D. stramonium* was included in the pharmacopeias of Mexico, the United States, as well as many other countries in America and Europe. Various respiratory diseases, especially asthma, have been traditionally treated using this plant by means of various preparations including powders and cigarettes made from the leaves. The tropane alkaloids do indeed have bronchodilating effect, but are not safe to use, especially in children.[58]

All parts of the plant, especially the leaves, are employed for medicine and ritualistic purposes.[46] Cigarettes are made from the dried leaves and smoked to relieve the symptoms of asthma. Additionally, the seeds and leaves are dried and ground into a powder that is lit in order to produce fumes that are inhaled by patients with asthma. This practice is undertaken because the plant's phytochemical compounds function as natural bronchodilators.[55,57] Even though the plant does possess various medicinal properties, it is too dangerous (especially the seeds) to use as a home remedy.[55,59,60] Datura or Jimsonweed contains various tropane alkaloids, which have medicinal applications in rheumatism, as pain relievers, and for respiratory issues. Unfortunately, the hazards associated with this plant preclude its use in modern phytotherapy.[55,57,61,62]

10.5.13   LINDEN TREE (TILIA SPP.)

The flowers are taken as a tea, by themselves or combined with other herbs, for the treatment of insomnia and nervousness. The flowers are used in various forms in traditional medicine, especially decocted in water. The tea's main application in MTM is to promote a relaxing and restful sleep, since the flowers have a calming, and non-narcotic action. For this reason, the tea is used for anxiety. In addition, the flowers are employed to treat colds, and pyrexia (fever), due to their antiviral and diaphoretic action.[49] The plant is decocted in water along with quills of true cinnamon (*Cinnamomum verum*); the resulting tisane is strained, and taken preferably at night, for the treatment of cough.[3]

10.5.14   MOUNTAIN MINT (CUNILA LYTHRIFOLIA)

The whole plant is taken as a tea or the leaves can be boiled in water and the vapors are inhaled in order to treat cough and colds. Approximately, 10–15 g of the flower stems and leaves are decocted in three L of water for 2–4 min. The decoction is strained and allowed to cool. The tea can be taken 4 times per day for a period of 15–20 days or until the symptoms.[34]

10.5.15   DAMIANITA DAISY (CHRYSACTINIA MEXICANA)

The aerial part of the plant is decocted in water and taken as a tea for the treatment of cough and colds.[12] According to Mexican researchers, damianita daisy (commonly known as *Hierba de San Nicolás*) weed is among the most potent of 187 plants used in MTM for the treatment of respiratory conditions related to tuberculosis.[63]

10.5.16   HOLLOW OR PIPRANK PLANT (ARISTOLOCHIA TALISCANA)

The roots of *A. taliscana* contain neolignan compounds that have demonstrated important antimycobacterial effects for the potential treatment of tuberculosis and other respiratory diseases. However, no clinical trials are known that could corroborate its safety or efficacy.[63]

10.5.17   WORMSEED (DYSPHANIA GRAVEOLENS)

This very pungent species is closely related to the better-known species of wormseed (*D. ambroisoides*), with which it shares its anthelminthic (worm-expelling) properties. However, Mexican herbalists also regard this plant as more potent against various respiratory afflictions such as pneumonia, cough, and colds. In the state of Durango, the plant is combined with Mexican thistle or "hierba del sapo" (*Eryngium heterophyllum*) against cough.[64] Approximately, 10–15 g of the whole plant are decocted in 3 L of water for a period of 2–4 min. The resulting tea is taken 4 times a day for 15–20 days.[34] Both species of *Dysphania* contain ascaridol and therefore should be avoided during pregnancy and lactation.[64]

10.6 SUMMARY

A diverse array of medicinal plants, both native as well as introduced, have been used with success for centuries by traditional herbalists in Mexico in order to treat a plethora of health issues related to the respiratory tract. Although some have been researched as to their phytochemical constituents, many herbs still have not been properly studied with regard to their efficacy or safety. Even though they are "natural," some plants employed in traditional medicine may be toxic under certain circumstances.

KEYWORDS

- **respiratory diseases**
- **Mexico**
- **colt's foot**
- **Datura**
- **decoction**
- **epazote**
- **wormseed**

REFERENCES

1. Adame, J; Adame, H. *Plantas Curativas del Noreste Mexicano* (Curative Plants in the North-East of Mexico). Ediciones Castillo, N. L.: Monterrey, México, 2000; p 53.
2. Aguilar, A; Camacho, J. R.; Chino, S.; Jacquez, P.; Lopez, M. E. *Plantas Medicinales del Herbario IMSS* (Medicinal Plants in the IMSS Herbal Park). Instituto Mexicano del Seguro Social (IMSS): México, DF, 1996; pp 59–61.
3. Allred, K; Ivey, R. *Flora Neomexicana* (Flora of New Mexico). Lulu Press: Raleigh, NC, 2012; Vol. 3, pp 166–167.
4. Argueta, A., Ed. *Plantas Medicinales de Uso Tradicional en la Ciudad de México* (Traditional Uses of Medicinal Plants in the Mexico City). UNAM: México, D.F., 2014; pp 67–68.
5. Argueta, A. *Atlas de la Medicina Tradicional Mexicana*, 4 tomos (Atlas of Traditional Mexican Plants: Four Volumes). Instituto Nacional Indigenista: México, D. F., 1994; p 317.
6. Barnes, P. M; Powell-Griner, E; McFann, K; Nahin, R. L. Complementary and Alternative Medicine Use among Adults: United States, 2002. *Adv. Data* **2004,** *27* (343), 1–19.

7. Berdonces, J. L. *Guía de Plantas Psicoactivas: Historia, Usos y Aplicaciones* (Guide to Psychoactive Plants: History, Use and Applications). Ediciones Invisibles: Barcelona, 2015; pp 79–81.

8. Bone, K; Mills, S. *Principles and Practice of Phytotherapy*, 2nd ed.; Churchill Livingstone: London, 2013; p 118.

9. Boumba, V. A.; Mitselou, A; Vougiouklakis, T. Fatal Poisoning from Ingestion of *Datura stramonium* Seeds. *Vet. Hum. Toxicol.* **2004,** *46* (2), 81–82.

10. Davidow, J. *Infusions of Healing.* Fireside Books: New York, 1999; p 119.

11. Dickinson, A. History and Overview of Dietary Supplement and Health Education Act (DSHEA). *Fitoterapia* **2011,** *82* (1), 5–10.

12. Duke, J; Bogenschutz-Godwin, M; Ottensen, R. *Duke's Handbook of Medicinal Plants of Latin-America.* CRC Press: Boca Raton, FL, 2009; p 213.

13. Eisenberg, D. M.; Davis, R. B.; Ettner, S. L.; Appel, S. Alternative Medicine use in the United States, 1990–1997. *J. Am. Med. Assoc.* **1998,** *280*, 1569–1575.

14. Emmart, E. W. *The Badianus Codex.* Johns Hopkins University Press: Baltimore, MD, 1942; p 315.

15. Funayama, A.; Cordell, G. *Alkaloids: A Treasury of and Medicines and Poisons.* Academic Press: New York, 2015; p 318.

16. Gómez-Cansino, R.; Guzmán-Gutiérrez, S. L.; Campos-Lara, M. G.; Espitia-Pinzón, C. I.; Reyes-Chilpa, R. Natural Compounds from Mexican Medicinal Plants as Potential Drug Leads for Anti-Tuberculosis Drugs. *Anais da Academia Brasileira das Ciencias*, **2017,** *89* (1), 31–43.

17. González-Elizondo, M; López-Enríquez, L *Plantas Medicinales del Estado de Durango* (Medicinal plants in the state of Durango). IPN/PROSIMA: México, D. F., 2004, p 59.

18. González, M. *Plantas Medicinales del Noreste de México* (Medicinal Plants in the North East of Mexico). IMSS-Grupo Vitro: Monterrey N. L., Mexico, 1998; p 72.

19. González-Stuart, A; Rivera, J. Comparison of Herbal Product Use in the Two Largest Border Communities between the US and Mexico. *Herb. Gram* **2009,** *81,* 58–65.

20. Hendrickson, B. *Border Medicine: Transcultural History of Mexican-American Curanderismo.* New York University Press: New York, 2014; pp 10–12, 80–82, 152–153.

21. Hosseinzadeh, S; Jafarikukhdan, A; Hosseini, A.; Armand, R. The Application of Medicinal Plants in Traditional and Modern Medicine: Review of *Thymus vulgaris. Int. J. Clin. Med.* **2015,** *6*, 635–642

22. Kane, C. W. *Herbal Medicine of the American Southwest.* Lincoln Town Press: Tucson, AZ, 2009; pp 184–186.

23. Krenzelok, E. P. Aspects of Datura Poisoning and Treatment. *Clin. Toxicol.* (Philadelphia), **2010,** *48* (2), 104–110.

24. León, A; Toscano, R. A.; Tortoriello, J.; Delgado, G. Phthalides and Other Constituents from *Ligusticum porter*: Sedative and Spasmolytic Activities of Some Natural Products and Derivatives. *Nat. Products Res.* **2011,** *25* (13), 1234–1242.

25. Levine, R. E.; Gaw, A. C. Culture-Bound Syndromes. *Psychiatric Clin. North Am.* **1995,** *18* (3), 523–536.

26. Linares, E.; Bye, R. A. Study of Four Medicinal Plant Complexes of Mexico and Adjacent United States. *J. Ethnopharmacol.* **1987,** *19* (2), 153–183.

27. Lozoya, X. *Xiuhpatli: Herba officinalis.* SSA/UNAM: México, D. F., 1999; pp 82–85.

28. Ma, L.; Gu, R.; Tang, L.; Chen, Z. E.; Di, R.; Long, C. Important Poisonous Plants in Tibetan Ethnomedicine. *Toxins* (Basel) **2015,** *7* (1), 138–155.

29. Mabberley, D. J. *Mabberley's Plant Book*, 4th ed.; Cambridge University Press: London, 2017; p 125.

30. Macouzet-Pacheco, M. V.; Estrada-Castillón, E.; Jiménez-Pérez, J.; Villarreal-Quintanilla, J. A; Herrera-Monsiváis, M. C. *Plantas medicinales de Miquihuana, Tamaulipas*. UANL: Monterrey, N. L., 2013; p 217.

31. Mai, N. T.; Cuc, N. T.; Anh, H. L. Steroidal Saponins from *Datura metel*. *Steroids* **2017**, *121*, 1–9.

32. Marcus, D. M. Dietary Supplements: What's in a Name? What's in the Bottle? *Drug Test Anal.* **2016**, *8* (3–4), 410–412.

33. Martínez, M. *Las Plantas Medicinales de México*. Editorial Botas: México, D.F., 1988; pp 36–37.

34. McGuffin, M.; Kartesz, J.; Leung, A.; Tucker, A. *Herbs of Commerce*, 2nd ed.; American Herbal Products Association: Silver Spring, MD, 2000; p 122.

35. McNeill, B.; Cervantes, J. M. *Latina/o Healing Practices: Mestizo and Indigenous Perspectives*. Routledge: New York, 2008; p 216.

36. Mendoza-Castelán, G.; Lugo-Pérez, R. *Plantas Medicinales en los Mercados de México* (Medicinal Plants in the Markets of Mexico). Universidad Autónoma Chapingo: Chapingo, México, 2011; pp 420–421.

37. Moerman, D. *Native American Ethnobotany*. Timber Press: Portland, OR, 1999; p 214.

38. Moore, M. *Los Remedios: Traditional Herbal Remedies of the Southwest*. Red Crane Books: Santa Fe, NM, 1995; pp 61–62.

39. Murphy, S. L.; Xu, J. Q.; Kochanek, K. D. Deaths: Final Data, 2010. *National Vital Statistics Reports*, Vol. 61; No. 4. National Center for Health Statistics: Hyattsville, MD, 2013. http://www.cdc.gov/nchs/data/nvsr/nvsr61/nvsr61_04.pdf (accessed Aug 29, 2017).

40. *Neuman, M. G. Hepat*otoxicity of Pyrrolizidine Alkaloids. *J. Pharm. Pharm. Sci.* **2015**, *18* (4), 825–843.

41. Neumann, K. H.; A. Kumar, A.; Imani, Y. J. *Plant Cell and Tissue Culture. A Tool in Biotechnology, Principles and Practices*. Springer-Verlag: Berlin, 2009; p 181.

42. Pan American Health Organization (PAHO). Mortality—Principle Causes of Death in México, 2010. http://www.paho.org/data/index.php/es/mnu-mortalidad/principales-causas-de-muerte.html (accessed Aug 27, 2017).

43. Pennington, T. D.; Reynel, C.; Daza, A. *Illustrated Guide to the Trees of Peru*. David Hunt: Sherborne, England, 2004; pp 147–148.

44. Porter, R. S.; Bode, R. F. Review of the Antiviral Properties of Black Elder (*Sambucus nigra* L.) Products. *Phytother. Res.* **2017**, *31* (4), 533–554.

45. Quattrocchi, U. *World Dictionary of Medicinal and Poisonous Plants,* Vol. 5; CRC Press: Boca Raton, FL, 2012; p 736.

46. Ratsch, C. *Encyclopedia of Psychoactive Plants*. Park Street Press: Rochester, V T, 2005.

47. Redzowski, J. *Vegetación de México* (Vegetation of Mexico). Limusa: México, D. F., 1978, p 119.

48. Rivera J. O.; Ortiz, M.; González-Stuart, A.; Hughes, H. Bi-National Evaluation of Herbal Product use on the United States/México Border. *J. Herb. Pharmacother.* **2007**, *7* (3–4), 91–106.

49. Rivera, J. O.; Hughes, H. W.; Gonzalez-Stuart, A. Herbals and Asthma: Usage Patterns Among a Border Population. *Ann. Pharmacother.* **2004**, *38* (2), 220–225.

50. Rodríguez-Ramos, F.; Navarrete, A. Solving the Confusion of Gnaphaliin Structure: Gnaphaliin A and Gnaphaliin B Identified as Active Principles of *Gnaphalium Liebmannii* with Tracheal Smooth Muscle Relaxant Properties. *J. Nat. Prod.* **2009,** *72* (6), 1061–1064.

51. Sandoval, A. *Homegrown Healing: Traditional Home Remedies from Mexico.* Berkley Books: New York, 1996, p 118.

52. Slater, J.; López-Terrada, M.; Pardo-Tomás, J. *Medical Cultures of the Early Modern Spanish Empire.* Ashgate Publishing: Surrey, England, 2014; pp 8–10.

53. Stillman, A. S. Hepatic Veno-Occlusive Disease Due to Pyrrolizidine (Senecio) Poisoning in Arizona. *Gastroenterology* **1977,** *73* (2), 349–352.

54. Stubbendieck, J.; Hatch, S.; Neill, C. *North American Wild Land Plants,* 3rd ed.; University of Nebraska Press, Lincoln, NE: Lincoln, NE, 2017; p 142.

55. Taddei-Bringas, G. A.; Santillana-Macedo, M. A.; Romero-Cancio, J. A.; Romero-Tellez, M. B. Acceptance and Use of Medicinal Plants in Family Medicine. *Salud Pública México* **1999,** *41,* 216–220.

56. Tola, J.; Infiesta, E. *Principales Plantas Medicinales de Europa y América* (Principle Medicine Plants of Europe and America). Robin Book: Barcelona, 2000; p 238.

57. Torres, E. *Healing with Herbs and Rituals.* University of New Mexico Press: Albuquerque, NM, 2006; p 98.

58. Trujillo, W. A.; Sorenson, W. R. Determination of Ephedrine Alkaloids in Dietary Supplements and Botanicals by Liquid Chromatography/Tandem Mass Spectrometry: Collaborative Study. *J Assoc. Off. Anal. Chem. Int.* **2003,** *86* (4), 657–668.

59. Vega-Ávila, E.; Espejo-Serna, A.; Alarcon-Aguilar, F.; Velasco-Lezama, Y. R. Cytotoxic Cativity of Four Mexican Medicinal Plants. *Proc. West Pharmacol. Soc.* **2009,** *52,* 78–82.

60. Villaseñor, J. L.; Espinosa-Garcia, F. J. The Alien Flowering Plants of Mexico. *Divers. Distrib.* **2004,** *10,* 113–123.

61. White, R. *Elsevier's Dictionary of Plant Names of North America Including Mexico.* Elsevier: Amsterdam, 2003; p 90.

62. Wiley, A.; Allen, J. *Medical Anthropology,* 2nd ed.; Oxford University Press: London, 2013; pp 345–350.

63. Yardley, K. *The Good Living Guide to Herbal Remedies.* Good Books: New York, 2016; pp 150–151.

64. Zheng, X.; Wang, W.; Piao, H.; Xu, W.; Shi, H.; Zhao, C. The Genus *Gnaphalium* L. (Compositae): Phytochemical and Pharmacological Characteristics. *Molecules* **2013,** *18* (7), 8298–8318.

PART IV
Novel Applications of Plants

CHAPTER 11

BIOACTIVE COMPOUNDS IN COFFEE: HEALTH BENEFITS OF MACRONUTRIENTS AND MICRONUTRIENTS

ROSA LELYANA

ABSTRACT

Coffee has two commercial species, that is, *Arabica* and *Robusta* coffee. Coffee offers many health benefits that are affected by determinant factors. Coffee has various bioactive components, such as caffeine, alcoholic diterpene (caffestol and kahweol), and polyphenol in the form of chloregenic acid. The structure of polyphenol has several hydroxyl groups and aromatic rings. Arabica coffee has more lipids. Robusta has more caffeine and sucrose, such as, polyphenolic antioxidant, chlorogenic acid, and its derivatives. Coffee served through filtration process (*filtered coffee/brewed coffee*) does not increase the serum lipid levels; whereas coffee in the form of boiled coffee has more increase in serum lipid level. Coffee components, the methods of serving the coffee and chemical structure of biocompounds determine the health benefits of coffee.

11.1 INTRODUCTION

Coffee is one of favorite drinks for many persons in the world and it offers several health benefits.[16-25] Two popular coffee species are *Arabica* coffee and *Robusta* coffee. Arabica feels more likeable and dominate 80% of trade worldwide than Robusta.[2] Robusta coffee beans have more polyphenolic acid than Arabica coffee.[17]

This review discusses the determinant factors that influence the health benefits of coffee.

11.2 BIOCOMPONENTS OF COFFEE FOR HEALTH BENEFITS

11.2.1 POLYPHENOLS IN COFFEE

Polyphenols are available in a variety of plants including coffee.[16] Polyphenol structures consist of hydroxyl groups and aromatic rings.[32] Chlorogenic acid reduces hepatic triglyceride levels. Chlorogenic acid inhibits glucose-6-hepatic-phosphatase (G6PD) thus limiting the enzyme related to gluconeogenesis.[19] Roasting the coffee will reduce the chlorogenic acid percentage in coffee. Green coffee beans are rich in chlorogenic acid and are linked to the effects of hypotension, beneficial increase of fat metabolism in the liver. The research studies on coffee beans using the sesamin and orlistat reagents as reference compounds showed better fat reduction if the provision of coffee drink is in the form of whole food, rather than the form of active substances such as chlorogenic acid or caffeine alone because chlorogenic acid will increase the metabolism of fat in the liver.[37] In addition, the effect of increased blood pressure from caffeine will be more influential when caffeine is consumed alone.[12]

11.2.2 OTHER BIOACTIVE COMPOUNDS CONTAINED IN COFFEE[29]

- 1.5-quinolactones is useful for reducing glucose in adenosine receptors of the brain.
- 5-caffeoxylquinic acid is a major chlorogenic acid in coffee that affects the decreased risk of developing *diabetes mellitus.*
- Caffeic acid lowers Nf-KB levels, which decrease the pro-inflammatory response.[21] Caffeic acid is a class of hydroxy-cinnamic acids.
- Lignans are phytoestrogens that have fat-soluble antioxidant activity to fight against free radicals. Secoisolarisiresinol diglucoside (a dietary antioxidant lignan in coffee) helps to reduce the risk of developing *diabetes mellitus* in streptozotocin-induced mice.[12]
- Melanoidin is a brown polymer that has antioxidant activity, formed by Mallard reaction in coffee roasting.

- Pyroglutamate.
- Tannin belongs to an ester family, including one of phenolic compounds in coffee beans. Tannin content is reduced when soaked in water and alkaline solution.
- PGA is an intermediate of glutathione metabolism. PGA levels rise 10-folds after processing; therefore, there are higher levels in instant coffee than raw coffee. Beans are useful to treat neurologic cases, such as, Alzheimer's because it increases *Ach*.[1]
- Trigonelline (N-methylnicotinic acid) has hypoglycemic effect in aloxan-treated diabetic rats.[12] Trigonelline is an intermediate metabolism of vitamin B_3, one of many alkaloids contained raw coffee beans. Trigonelline is a therapeutic coffee compound.[1]

11.3 CAFFEINE

Caffeine is one of the alkaloids; is common sense of the methylxanthin family; and a stimulant of the central nervous system that is antagonistic to the action of adenosine. Caffeine is a competitive antagonist of the adenosine receptor. Adenosine works as a depressant that bears a resemblance to the caffeine molecule so it can be occupied by alkaloids. Caffeine is often used as a drug formulation capable of suppressing the depressive effects of other drugs.[28]

The content of caffeine in beverages varies depending on the size of the product type and the method of processing. The positive[21] and negative effects of coffee consumption on health depend on the level of caffeine embedded in the drink. The upper limit of caffeine consumption for adult health is 300–500 mg per day, while for pregnant women is between 150 mg and 200 mg per day and children under 50 mg per day. Caffeine levels that exceed 700 mg per day will have negative effects on our health.[34]

Caffeine is a safe thermogenic agent for weight control. The short-term lethal dose of caffeine is about 5–10 g per day [in IV or oral]. Five to ten gram of caffeine per day is equivalent to consumption of 75 cups of coffee, 125 cups of tea, 200 cola drinks. Only slight changes were not significant in blood pressure and pulse after consuming 100 mg and 200 mg of caffeine. Furthermore, at a dose of 400 mg, caffeine will significantly increase the systolic and diastolic blood pressure of about 6.3 mmHg. A dose of 250 mg of caffeine taken orally will increase blood pressure by 10 mmHg after 1 h of caffeine consumption. A dose of 400 mg of caffeine will also significantly increase the side effects of palpitation, anxiety, dizziness, lack of fit, ringing ears. According to previous studies on the chronic effects of caffeine consumption of 150 mg per day for 7 days, caffeine tolerance occurs after 1–4 days. Therefore, there are no long-term effects of caffeine on blood pressure, heart rate, plasma renin activity. In addition, there was no significant change in heart rate after consuming 4 mg of caffeine per kg of body weight as much as 5 times per day. Thus, caffeine is relatively safe, although in acute consumption it will alter cardiovascular variables and chronic consumption has little or no effect on health.[9]

Alkaloids are able to decrease the glucose transport and inhibit α-glucosidase, an enzyme that hydrolyzes maltose into glucose (thus helping to decrease glucose levels).[3] Based on the result of the study of three brands of packaged coffee powder sold in mini market, Semarang contains caffeine varying from 8.310 to 20.429 mg/kg.[19]

Caffeine (1,3,7-trimethylxanthin) is one of the most commonly consumed psycho-stimulants. The effects of caffeine are mediated at several targets, such as, cAMP phosphodiesterase, phosphatidylinositol-3 kinase, and adenosine receptors. Pharmacological stimulation of A2AR agonists results in inhibition of proliferation, cytokine production, and T cell cytotoxicity, and activation of monocytes and granulocytes. The anti-inflammatory effects of A2AR agonists have been demonstrated in vivo, including ischemia/reperfusion injury, airway inflammation, and acute T cell-dependent hepatitis. The effect of caffeine protection by suppressing hepatic damage is due to the A2AR-independent mechanism and largely by inhibiting cAMP phosphodiesterase.[31]

Caffeine suppresses fat absorption. Caffeine promotes lipolysis of fats, increases energy expenditure, encourages sugar consumption, and increases blood epinephrine hormone in rats. Consumption of caffeine before physical exercise will encourage ventilation and improve lipolysis.[31,37] Coffee caffeine as a lipoytic food component will improve endurance in animal studies of rats.[35]

Caffeine is diuretic and has a pressor and ergogenic effect, potentially improving physical work associated with thermogenesis, lipolysis, fat oxidation, and glucose metabolism.[12]

The pressor effect of caffeine arises after consuming caffeine and caffeine undergoing digestive processes in the gastrointestinal tract. The effects of pressor may increase the risk of cardiovascular disease especially in someone with a history of hypertension or risk factors for cardiovascular disease.[12]

11.3.1   EFFECTS OF CAFFEINE ON RESPONSE OF NEURAL SYSTEM

The sympathetic nervous system plays an important role in the regulation of the cardiovascular system. Under stress, borderline hypertensive patients show increased sympathetic nervous reactivity, as do individuals with *normotensi* from parents with a history of hypertension.[30] When coffee or caffeine is given acutely to nonhabitual coffee drinkers with the same dose of caffeine plasma, there will be an increase in sympathetic nerve activity and blood pressure. In contrast, in habitual coffee drinkers there is a reduction in the response to increased blood pressure despite an increase in sympathetic nerve activity.[39] Caffeine can lower blood pressure.[42]

11.3.2   CAFFEINE METABOLISM

The metabolism of caffeine is affected by sex differences resulting in different physical activity responses between men and women. Caffeine metabolism depends on the activity of the P450 cytochrome, especially the CYP1A2 isoform. Women are more likely to express lack of CYP1A2 activity than men so that caffeine half-life is greater in women where longer stimulation actions are due to elevated levels of plasma caffeine over the long term. The difference is also influenced by steroid hormone levels, which in the follicular phase show a similar caffeine response among men, but women's luteal phase is less responsive than men. It affects the increased physical activity in women than men due to caffeine consumption.[36]

The high physical activity, associated with high androgen levels and the emergence of anti-atherogenic signals, resulted in a lack of aortic classification in elderly women with peripheral obesity. Women with peripheral obesity will show elevated levels of DHEA and ASD according to the degree of physical activity. Physical activity is considered as one of the main factors in the management of body weight. Physical exercise will reduce visceral fat and fat, increase energy expenditure [energy expenditure], improves glycemic control. Exercise/physical activity will affect the metabolic changes in some organs of the body, such as, the hypothalamus, liver, and adipose tissue.[38]

11.3.3 EFFECTS OF DECAFFEINATED COFFEE IN OUR BODY

Decaffeinated coffee increases the blood pressure of sympathetic nerve activity in nonhabitual coffee drinkers, thus providing an understanding that there are other components in coffee that activate the cardiovascular system.[39]

11.4 FIBER

Coffee drink contains high fiber. Based on the results of the study, three brands of coffee powder in the mini market of Semarang city contain about 16% fiber.[19] The fiber content in the coffee powder will be able to inhibit the activity of alpha amylase, an enzyme that helps the body convert food into energy and catalyze starch hydrolysis into sugars, thereby reducing glucose absorption rate. Instant coffee powder that has been roasted and brewed at 100°C will contain soluble fiber. Galactomanan fiber content and arabino-galactan are found in brewed coffee. Conversely, hot water contains only nonsoluble fiber in green coffee beans[33] not yet roasted and not brewed.[16,17,19]

Polysaccharides in the coffee powder can increase insulin levels, reduce blood glucose levels, and improve glucose tolerance. Approximately 50% of polysaccharides are contained in three brands of packaged coffee powder sold in Semarang mini market.[19]

11.5 HEALTH BENEFITS OF MACRONUTRIENTS IN COFFEE

11.5.1 PROTEIN—TRYPTOPHAN

Tryptophan is a protein that is found in green beans. Tryptophan content will affect the increase in serotonin levels.[6] Serotonin is an appetite controller. Serotonin function will reduce the intake of food by a person. Conversely, decreased serotonin levels will increase appetite.[4] Patients with depression will experience deficiency of serotonin levels so that more food is intaken than energy expenditure, resulting in visceral fat accumulation and *proinflamasi* cytokines.[26,27]

The low levels of tryptophan plasma with low also tryptophan ratio to other neutral amino acids in obesity. Plasma tryptophan levels are estimated from brain tryptophan levels used and how much serotonin is produced by the brain. Tryptophan levels were lower in obesity when compared with nonobese person.[40]

11.5.2   CARBOHYDRATES

The content of carbohydrates in the food and amount of beverage consumption affects the development of an `obese person. Coffee drink contains carbohydrates. Not all carbohydrates are similar in chemical structure: not just the number of different carbon chains, but the speed of metabolism also plays an important role in the prevention of high-carbohydrate dietary problems. The nutritive content of coffee is in the form of macronutrients. Carbohydrates are the main component of coffee beans. About 50% of the dry weight of raw coffee beans is carbohydrates.[19,41]

Carbohydrates contribute to the taste of coffee. Carbohydrates undergo complex changes, for example, react with amino acids during Maillard reaction when undergoing roasting. Carbohydrates work as a scent concealer, a stabilizer of froth, and affect the viscosity of coffee beverage. Carbohydrates are also the assessors of the authenticity of instant coffee drink.[41] High-fiber carbohydrates have lower glycemic index levels that will be digested more slowly and reduce postprandial hyperglycemia.[18] Carbohydrates are able to increase plasma levels of Tryptophan 5HT precursor.[27]

11.5.3   LIPIDS

Diterpenoid is the oil portion in the coffee and levels depend on how the coffee is served. The results of the study showed that antioxidant content, when coffee is brewed with hot water, is higher in Robusta coffee compared to other coffee types in Semarang market.[17] In vitro results showed that the diterpenoid reduced the activity of LDL receptors in cells.[5]

11.6   EFFECTS OF OTHER NUTRIENT COMPONENTS IN COFFEE ON OUR HEALTH

11.6.1   VITAMINS

Niacin content of about 2–80 mg/100 g of coffee depends on the authenticity of coffee beans, roasting, and preparation process of coffee beans. Niacin is present in coffee and is formed during the roasting process of coffee beans.[10]

11.6.2   MINERALS

Coffee consumption provides an 8% daily intake of Cr and is a substantial source of Mg with an average of 63.7 per cup (100 mL). In fact, Mg excretion is positively correlated with Mg intake and coffee consumption. Mg content in coffee can affect the increase of insulin sensitivity so that long-term coffee consumption may decrease the risk of type-2 *diabetes mellitus*.[12] Coffee can negate toxic metals such as Pb from the water content.[10]

11.7   SERVING METHODS OF COFFEE

Coffee drinking will increase or decrease the boost immune system.[24] Serving coffee components will influence the activity of antioxidant components of coffee. Robusta coffee is mostly used in the form of instant coffee and as filler for roasting and mixing. The roasting process causes water loss from coffee beans as well as degradation of various compounds including polyphenolic antioxidants.[33]

Decaffeinated coffee has about 3–4 mg of caffeine per cup of coffee. Ground coffee, drip coffee, and instant coffee has about 65–110 mg of caffeine per cup of coffee. Brewed coffee has about 8–135 mg of caffeine per cup of coffee.[13]

11.8   CHEMICAL STRUCTURE

Coffee drinking could decrease the uric acid blood level.[7,8,14,17,23] There are higher phenolic acid total of pure Robusta coffee than Arabica coffee beans,[17] which has beneficial effect as antioxidant (due to chlorogenic acid) and inhibit xanthin oxidase,[11] thus decreasing the uric acid blood level. Caffeine's chemical structure is 1,3,7-trimethylxanthine with inclusion of xanthine molecule[15] so that chemical structure of caffeine resembles with xanthine. This resemblance in chemical structure between caffeine and xanthine has potentiality for inhibiting xanthine oxidase that will decrease uric acid blood level.[17]

11.9   SUMMARY

This chapter focuses on health benefits due to coffee drink and its components, methods of serving the coffee, and chemical structure of caffeine.

KEYWORDS

- **antioxidant**
- **caffeine**
- **macronutrients**
- **micronutrients**
- **polyphenol**

REFERENCES

1. Akashi, I.; Kagami, K.; Hirano, T.; Oka, K. Protective Effects of Coffee Derived Compounds on Lipopolysaccharide/D-Galactosamine Induced Acute Liver Injury in rats. *J. Pharm. Pharmacol.* **2009,** *61* (4), 473–478.
2. Baldwin, J. *Arabica* versus *Robusta*: No Contest. *The Atlantic Daily*, 2009; p 1. https://www.theatlantic.com/health/archive/2009/06/arabica-vs-robusta-no-contest/19780/ (accessed on Oct 11, 2017).
3. Battram, D. S.; Arthur, R.; Weekes, A. The Glucose Intolerance Induced by Caffeinated Coffee Ingestion is Less Pronounced than that Due to Alkaloid Caffeire in Men. *J. Nutr.* **2006,** *136*, 1276–1280.
4. Breum, L.; Rasmussen, M. H.; Hilsted, J.; Fernstrom, J. D. Twenty Four Hour Plasma Trypytophan Concentrations and Ratios are Below Normal in Obese Subjects and are Not Normalized by Substantial Weight. *Am. J. Clin. Nutr.* **2003,** *77* (5), 1112–1118.
5. Cardenas, C.; Quesada, A. R.; Medina, M. A. Antiangiogenic and Antiinflammatory Properties of Kahweol, a Coffee Diterpene. *PLoS One* **2011,** *6* (8), E-article: 23407.
6. Cavaliere, H.; Medeiros-Neto, G. The Anorectic Effect of Increasing Doses of L-Tryptophan in Obese Patients. *Eat. Weight Disord.: EWD* **1997,** *2* (4), 211–215.
7. Choi, H. K; Curhan, G. Coffee, Tea and Caffeine Consumption and Serum Uric Level—Third National Health and Nutrition Examination Survey. *Arthritis + Rheuma* **2007,** *57* (5), 816–821.
8. Choi, H. K.; Willett, W.; Curhan, G. Coffee Consumption and Risk of Incident Gout in Men: A Prospective Study. *Arthritis + Rheuma* **2007,** *56* (6), 2049–2055.
9. Diepvens, K.; Westerterp, K. R.; Plantenga, M. S. W. Obesity and Thermogenesis Related to the Consumption of Caffeine, Ephedrine, Capsaicin, and Green Tea. *Am. J. Physiol. Regul. Integr. Comp. Physiol.* **2007,** *292*, R77–R85.
10. Dorea, J. G.; da Costa, T. H. M. Is Coffee a Functional Food? *Br. J. Nutr.* **2005,** *93* (6), 773–782.
11. Farah, A.; Donangelo, M. C. Phenolic Compounds in Coffee with Antioxidant Inhibit Xanthin Oxidase. *Revista Brasileira de Fisiologia Vegetal* **2006,** *18*, 23–26.
12. Greenberg, J. A.; Boozer, C. N.; Geliebter, A. Coffee, Diabetes, and Weight Control. *Am. J. Clin. Nutr.* **2006,** *84*, 682–693.

13. James, J. E. Caffeine and Health. In *Progress in Clinical and Biological Research*; Spiller, G. A., Liss, A. R., Eds.; Academic Press: New York; Vol. 158; 1984; p 135.

14. Kiyohara, C.; Kono, S.; Honjo, S.; Todoroki, I.; Sakurai, Y. Inverse Association Between Coffee Drinking and Serum Uric Acid Concentration in Middle Age Japanese Males. *Br. J. Nutr.* **1999**, *82* (2), 125–130.

15. Lanchane, M. P. The Pharmacology and Toxicology of Caffeine. *J. Food Safety* **1982**, *4* (2), 71–112.

16. Lelyana, R. *Kandungan polifenol* (Muwarni R), *dan kafein* (Suzzeri M) *dalam kopi*: *Pengaruh Kopi Terhadap Kadar Asam Urat* [Polyphenol content (Muwarni R), and Caffeine (Suzzeri M) in Coffee: Influence of Coffee against Uric Acid Level]. Master of Biomedical Thesis, Post Graduate Program at Sarjana Diponegoro University, Semarang, Indonesia, 2008; p 125.

17. Lelyana, R.; Cahyono, B. Total Phenolic Acid Content in Some Commercial Brands of Coffee from Indonesia. *J. Med. Plant Herb. Ther. Res.* **2015**, *3* (2), 27–29.

18. Lelyana, R. *Efek polifenol dan kafein terhadap kadar asam urat dalam mencegah sindrom metabolic* (Effects of Polyphenols and Caffeine on Uric Acid Levels in Preventing Metabolic Syndrome). *Keep Healthy and Young But How?,* Proceeding of NASWAAM-PT. Blesslink Rema on Udayana University: Denpasar, Bali, 2014; p 8.

19. Lelyana, R. The Role of Coffee's Content for Preventing Diabetes Mellitus Risk Factor Incidence. In *Book of Proceeding & Abstract Book*; 9th Symposium on Nutri Indonesia: Breaking the Boundaries to Optimize the Benefits for Patients; Universitas Indonesia: Jakarta, 2014; p 2.

20. Lelyana, R. Influence of Coffee on the Mycobacterium TBC Antibacterial Response—Review. At International Symposium Drug and Vaccine Development in Post MDG's Era; Gajah Mada University: Yogyakarta, Indonesia, 2014; p 10.

21. Lelyana, R. Anticarcinogenic activity of coffee consumption by suppressing overexpression of adipose tissue inflammatory responding in obesity people. At the 3rd International Seminar on Chemistry; Bandung, Indonesia: Universitas Padjadjaran; November 20-21, **2014**; pages 6.

22. Lelyana, R.; Wijayahadi, N. The Influence of Coffee Consumption to Decrease Uric Acid Level: An Experimental Study on Hyperuricemia Wistar strain Rats. *Int. J. Curr. Res.* **2007**, *7*, 19147–19153.

23. Lelyana, R. Effect of Coffee on Daily Consumption on Uric Acid Level and Body Weight to Prevent Metabolic Syndrome. *J. Nanomed. Nanotechnol.* **2016**, *7*, 400–405.

24. Lelyana, R. Underlying Mechanism of Coffee as Inhibitor Adipogenesis for Complementary Medicine Use in Obesity. *J. Nanomed. Nanotechnol.* **2017**, *8*, 425–429.

25. Lelyana, R. Drinking Coffee for Life Style and Maintenance Immune System. *J. Nanomed. Nanotechnol.* **2017**, *8*, 146–150.

26. Mangge, H.; Summers, K. L.; Meinitzer, A.; Zelzer, S. Obesity-Related Dysregulation of the Tryptophan-Kynurenine Metabolism: Role of Age and Parameters of the Metabolic Syndrome. *Obesity (Silver Spring)* **2014**, *22* (1), 195–201.

27. Markus, C. R. Effects of Carbohydrate on Brain Tryptophan Availability and Stress Performance. *Biol. Psychol.* **2007**, *76* (1–2), 83–90.

28. Mazzafera, P. Degradation of Caffeine by Microorganisms and Potential Use of Decaffeinated Coffee Husk and Pulp in Animal Feeding. *Scientia Agricola* **2002**, *59* (4), 815–821.

29. Nardini, M.; Cirillo, E.; Natella, F.; Mencarelli, D.; Comisso, A.; Scaccini, C. Detection of Bound Phenolic Acids: Prevention by Ascorbic Acid and Ethylenediaminetetraacetic Acid of Degradation of Phenolic Acids During Alkaline Hydrolysis. *Food Chem.* **2002,** *19*, 119–124.

30. Nooll, G.; Wenzel, R. R.; Scheneider, M.; Oesch, V.; Binggeli, C.; Shaw, S.; Weidmann, P.; Lüscher, T. F. Increased Activation of Sympathetic Nervous System and Endothelin by Mental Stress in Normotensive Offspring of Hypertensive Parents. *Circulation* **1996,** *93* (5), 866–869.

31. Ohta, A.; Lukashev, D.; Jackson, E. K.; Fredholm, B. B.; Sitkovsky, M. The 1,3,7-Trimethylxanthine (Caffeine) May Exacerbate Acute Inflammatory Liver Injury by Weakening the Physiological Immunosuppressive Mechanism. *J. Immunol.* **2007,** *179*, 7431–7438.

32. Omoigui, S. Interleukin 6 Inflammation Pathway from Cholesterol to Aging-Role of Statins, Biphosphonate, and Plant Polyphenols in Aging and Age Related Diseases. *Immun. Ageing* **2007,** *4*, 1–9.

33. Onakpoya, I.; Terry, R.; Ernst, E. The Use of Green Coffee Extract as Weight Loss Supplement: A Systemic Review and Meta-Analysis of Randomized Clinical Trials. *Gastroenterol. Res. Pract.* **2011,** *2011*, E-article ID: 382852; DOI: 10.1155/2011/382852.

34. Phan, T. T. D.; Kuban, V.; Kracmar, S. Determination of Caffeine Contents of Coffee Brands in the Vietnamese Market. *J. Microbiol. Biotechnol. Food Sci.* **2012,** *1*, 995–1002.

35. Ryu, S.; Choi, S. K.; Joung, S. S.; Suh, H. Caffein as a Lipolytic Food Component Increases Endurance Performance in Rats and Athletes. *J. Nutr. Sci. Vitaminol.* **2001,** *47* (2), 139–146.

36. Schrader, P.; Panek, L. M.; Temple, J. L. Acute and Chronic Caffeine Administration Increases Physical Activity in Sedentary Adults. *Nutr. Res.* **2013,** *33* (6), 457–463.

37. Shimoda, H.; Seki, E.; Aitani, M. Inhibitory Effect of Green Coffee Bean Extract on Fat Accumulation and Body Weight Gain in Mice. *BMC Complement. Altern. Med.* **2006,** *6*, 9–15.

38. Traub, R. H.; Tanko, L. B. Higher Physical Activity is Associate with Increased Androgens, Low Interleukin 6 and less Aortic Calcification in Peripheral Obese Elderly Women. *J. Endocrinol.* **2008,** *199* (1), 61–68.

39. Sudano, I.; Spieker, L.; Binggeli, C.; Ruschitzka, F.; Luscher, T. F.; Noll, G.; Corti, R. Coffee Blunts Mental Stress-Induced Blood Pressure Increase in Habitual But Not in Nonhabitual Coffee Drinkers. *Hypertension* **2005,** *4* (3), 521–526.

40. Sugrue, M. F. Neuropharmacology of Drugs Affecting Food Intake. *Pharmacol. Therapeutics* **1987,** *32*, 145–182.

41. Thermo Fischer Scientific Inc. Carbohydrate in Coffee; 2012; AOAC Method 995, 13 Versus a New Fast Ion Chromatography. http://www.thermoscientific.com/dionex (accessed Jan 3, 2012).

42. Yeh, T. C.; Liu, C. P.; Cheng, W. H.; Chen, B. R. Caffeine Intake Improves Fructose Induced Hypertension and Insulin Resistance by Enhancing Control Insulin Signaling. *Hypertension* **2014,** *63* (3), 535–541.

CHAPTER 12

IN-VITRO ANTIDERMATOPHYTIC BIOACTIVITY OF PEEL EXTRACTS OF RED BANANA (*MUSA ACUMINATE*) AND COMMON BANANA (*MUSA PARADISICA*)

SHIVAKUMAR SINGH POLICEPATEL, PAVANKUMAR PINDI, and VIDYASAGAR GUNAGAMBHERE MANIKRAO

ABSTRACT

In-vitro antidermatophytic activity of high interpolar methanolic fruit peel extracts of red banana and common banana varieties were evaluated in this chapter. About 98% of methanolic soxhlet extract of fruit peels of red banana (*Musa acuminate*) and common banana (*Musa paradisica*) varieties were evaluated against dermatophytic fungi, namely, *Trichophyton rubrum*, *Trichophyton tonsurans*, *Trichophyton mentagrophytes*, *Micro sporium gypseum*, and *Candida albicans*; and bacteria, namely, *Staphylococcus aureus*, *Psudomonas aeruginosa*, *Bacillus subtilis*, and *Escherichia coli*. Utmost activity was observed in red banana extract. The minimum inhibitory concentrations, minimum fungicidal concentrations, and minimum bactericidal concentrations were determined against selected test strains. This study provides a basis for the isolation and purification of antidermatophytic active molecules from the fruit peels of red banana variety.

12.1 INTRODUCTION

Therapeutic plants are innate sources of active molecules that can be used against numerous illnesses.[9] The present challenges involves discovery of therapeutic plants with elevated natural commotion, squat toxicity, and with least cost.[1] Several studies have utilized plants to research potential against microbial therapeutics and with beneficial activities of the selected plant parts. However, studies are not sufficient to wrap the global biodiversity and conventional uses of therapeutic vegetation.

WHO indicates that 80% global revenue from wages of individuals in budding country is dependent on therapeutic vegetation.[2] The health cost of using drugs is much higher compared to use of plant-based medicines that are safe and cost effective with least side-effects. There is urgent need for research on therapeutic plants and their metabolites.[11]

Investigators and healers of conventional medicines have documented health benefits of therapeutic plants to treat dermatological diseases. The ring worm is known as dermatophytes. The three main genera—*Microsporum*, *Trichophyton*, and *Epidermophyton*—are strongly allied plant—scientifically. Among these, *Microsporum* is a recurrent reason for ring worm of scalp and might furnish augment to ring worm in all parts of the carcass. While *Trichophyton* cause ringworm from the scalp as well as erstwhile areas of crust and nails. *Epidermophyton* is mainly accountable to ringworm affecting surface of the skin, hands, and feet; and has been found to interlace within the skin, and it does not assault tresses.[15] *Candida* species establishes in the gastro-intestinal tract, oral cavity, and vagina.[3]

Red banana is scientifically called as *Musa acuminate* L., whereas the common banana is scientifically named as *Musa paradisiaca* L., and both varieties belong to the family Musaceae. Banana is grown throughout the world as a resource of food and income for the cultivators. Banana fruit is also considered as the fourth most imperative food behind rice, wheat, and maize. Obviously, banana fruit has slender radioactive property due to its potassium content and small amount of isotope potassium. Parts of banana are used as medicine since ancient times. Conventionally, every part of the banana plant (such as fruit, stem juice, and flowers) have health benefits, such as antidiarrhea, anti-oxidant, dysentery, menorrhagia, anti-diabetes, antilithiatic, antitumor, antimutagenic, antibacterial, antifungal, heptatoprotective, hypocholesterolemic, antimenorrhagia, antihelminthic, anti-ulcerogenic, hair growth promotor, wound healing, and inflammation, pain, and snakebite.[5–8,10,14,19]

The literature review indicated no reports available for a comparative evaluation of antidermatophytic activity of fruit peel of banana using high-interpolar methanolic extract. The present chapter focuses on phytochemical and antidermatophytic activity of red banana (*Musa acuminate*) and common banana (*Musa paradisiac*) varieties.

12.2 METHODS AND MATERIALS

12.2.1 COMPILATION OF FRUIT PEELS

Common banana was collected from Mahabubnagar of Telangana state and red banana was collected from Kamalapur of Karnataka state in India. Both were identified with Flora of Gulbarga district and Mahabubnagar in Telangana.[16] The coupon samples (HPU-54) were deposited into the herbarium of Department of Botany at Palamuru University in Telangana State. Gathered fruit peel resources were washed with distilled water to eliminate the contaminants and the samples were shade dehydrated using a surface of the paper at ambient temperature for 20 days.

12.2.2 EXTRACTION OF HIGH INTERPOLAR METHANOLIC EXTRACT OF FRUIT PEEL BY USING SOXHLET APPARATUS

The two varieties of fruit peel materials after shade-drying were powdered using grinder and 30 g of dry powder of fruit peels were weighed. The extraction was conducted using high interpolar methanolic solvent with a Soxhlet apparatus for 48 h. The extracts were condensed and conserved in a refrigerant with sealed boxes for future use.

12.2.3 DERMATOPHYTIC MICROBES

Five dermatophytes fungi culture strains (namely, *Trichophyton rubrum*, *Trichophyton mentagrophytes*, *Trichophyton tonsurans*, *Microsporum gypseum*, and *Candida albicans*) and four dermatophytic bacteria cultures (such as, *Bacillus subtilis*, *Staphylococcus aureus*, *Pseudomonas aeruginosa*, and *Escherichia coli*) were obtained from the Department of Botany at Gulbarga University, Kalaburuge, to be used in this research study. The bacterial strains were developed using NB (Hi-media, M-002) at 37°C and maintained using NA slant at 4°C; and fungi strains were maintained on SDA at 28°C at PDA slope at 4°C.

12.2.4 EVALUATION OF ANTIDERMATOPHYTIC ACTIVITY USING A WELL DIFFUSION METHOD[12]

The assessment was conducted by cup plate method. About 18 to 22 mL of PDA media was poured into the germ-free petri-plates and authors had to wait for the sample to solidify. Fungi was observed after five days of old maintained strain. And fungi strains were suspended in the brackish solution (0.85% NaCl) and accustomed for their turbidity at 0.5 McFarland's standard. 1 mL of fungi strain was extended in excess of the medium via pure goblet purveyor. By means of Flame Sterilized Borer, obligatory concentration of in-sequence watered down extract (0.62 to 40 mg/mL) was supplemented to 20 µL in each well, whereas petri-plates were kept in the media for 1 h inside the refrigerant, and afterward were incubated at 37°C. Once incubated for 48 h, the petri-plates were studied further for the test. Width region of reticence was deliberated and articulated. Dimethyl formamide (DMF) was used as a negative indicator at the same time. The procedure was repeated for three times. The comparable process was followed for antidermophytic observations using NA media that was incubated at 37°C for 18 h.

12.2.5 MINIMUM INHIBITORY CONCENTRATION

The minimum inhibitory concentrations (MIC) were determined by using standard procedure from National Committee for Clinical Laboratory Standards (NCCLS).[13]

12.2.6 NUMERICAL INFORMATION INVESTIGATION

The experimentation was repeated three times. The statistical analysis was executed for analysis of variance (ANOVA), using STATISTICA 5.5 (Stat Soft Inc, Tulsa, OK, USA) software.

12.3 RESULTS AND DISCUSSION

For presence of secondary metabolites in extracts of fruit peels of two varieties of *Musa paradisica* was identified using standard procedure.[4] Tables 12.1 and 12.2 indicate occurrence of secondary metabolites, such

TABLE 12.1 Antidermatophytic Activity of High-Interpolar Methanolic Extract from Fruit Peels of *Musa Paradisica and Musa Acuminate*.

Selected medicinal plants & diff. conc. of crude extracts	Conc.	Reticence of zones								
		Test fungi					Test bacteria			
		Tr	Tt	Tm	Mg	Ca	Sa	Pa	Bs	Ec
mg/mL						Mm				
Common banana *fruit peel* (*Musa Paradisica*)	40	6.00 ±1.00	4.66 ±0.57	4.33 ±0.57	6.33 ±0.57	5.00 ±1.00	6.33 ±1.15	5.33 ±0.57	5.33 ±0.57	4.66 ±0.57
	20	5.00 ±00	-	-	5.33 ±0.57	-	6.00 ±0.00	5.00 ±0.00	-	5.00 ±1.00
	10	-	-	-	-	-	4.66 ±0.57	-	-	-
	5	-	-	-	-	-	-	-	-	-
	2.5	-	-	-	-	-	-	-	-	-
	1.25	-	-	-	-	-	-	-	-	-
	0.62	-	-	-	-	-	-	-	-	-
Red banana *fruit peels* (*Musa Acuminate*)	40	17.66 ±0.57	13.66 ±1.52	14.66 ±1.52	17.00 ±0.00	13.33 ±0.57	22.00 ±0.00	18.33 ±1.15	19.66 ±1.15	18.00 ±0.00
	20	15.66 ±1.52	11.33 ±0.57	12.66 ±0.57	16.33 ±1.15	11.33 ±1.15	17.66 ±1.52	16.00 ±0.00	16.66 ±0.57	14.33 ±0.57
	10	12.33 ±0.57	9.33 ±0.57	10.00 ±0.00	14.33 ±0.57	9.33 ±0.57	15.66 ±1.15	14.33 ±0.57	13.00 ±0.00	12.33 ±1.15
	5	10.00 ±0.00	7.00 ±0.00	9.33 ±0.57	12.33 ±1.15	7.33 ±0.57	13.00 ±0.00	12.33 ±1.15	12.33 ±0.57	10.00 ±0.00
	2.5	9.33 ±0.57	6.33 ±0.57	7.33 ±1.15	10.66 ±0.57	4.33 ±1.15	11.33 ±0.57	10.66 ±0.57	9.00 ±0.00	9.33 ±1.15

TABLE 12.1 (Continued)

Selected medicinal plants & diff. conc. of crude extracts	Conc.	Reticence of zones								
		Test fungi					Test bacteria			
		Tr	Tt	Tm	Mg	Ca	Sa	Pa	Bs	Ec
	1.25	7.33 ±1.15	NA	6.00 ±0.00	7.00 ±0.00	NA	10.66 ±1.15	9.00 ±0.00	6.33 ±0.57	7.00 ±0.00
	0.62	-	-	-	5.66 ±1.15	-	6.66 ±1.52	5.33 ±0.57	5.00 ±0.57	4.66 ±0.57
Ketoconazole (Kt)	2	25.00 ±1.15	27.33 ±0.57	27.00 ±0.00	25.66 ±1.52	24.00 ±0.00	-	-	-	-
Streptomycin sulphate (St)	2	-	-	-	-	-	30.33 ±1.15	29.33 ±0.00	27.00 ±0.00	27.33 ±0.00

Legend: Tr, Trichophyton rubrum; Tt, Trichophyton tonsurans; Tm, Trichophyton mentagrophytes; Mg, Microsporium gypseum; Ca, Candida albicans; Sa, Staphylococcus aureus; Pa, Psudomonas aeruginosa; Bs, Bacillus subtilis; Ec. Escherichia coli; NA, not active; Kt, Ketoconazole; St, Streptomycin Sulphate.

as, flavonoids, steroids, and saponins. High Interpolar Methanolic extract from fruit peels of red banana showed sturdy constructive retort to flavonoids, steroids, phenols, and tri-terpenoids. Whereas common banana peels extract demonstrated affirmative retort toward alkaloids and weak saponins (Table 12.1).

TABLE 12.2 Preliminary Phytochemical Tests for Secondary Metabolites in High-Interpolar Methanolic Extracts from Fruit Peels of *Musa Paradisica* and *Musa Acuminate*.

Secondary metabolites	Investigation	Fruit peels from	
		Common banana (*Musa paradisica*)	Red banana (*Musa acuminate*)
Alkaloids	Mr	PPP	A
	Drag.	P	A
	Wag.	A	A
Phenolics	H.Wtr	A	PPP
	Fecl$_2$	A	P
	Ellag.	P	PPP
Flavonoids	Fecl$_2$	A	PPP
	PbCooH	A	PP
	Shinod.	P	PPP
	Zinc/HCl test	P	P
Tannins	Gel.	A	P
Triterpenoids	Salk.	A	A
	Lib-Bur	P	A
Steroids	Salk.	A	A
	Lib-Bur	A	AP
Saponins	Foam.	A	PP
Glyco.	Kel-Killia	A	A
	H$_2$So$_4$	A	PP
	Moli.	A	P
	Glyco.	P	-

Legends: P, present; PPP, strongly present; APP, moderately present; A, absent.

Antidermatophytic analysis provides the direction of establishment for each of fungi and bacteria strain, in case of red banana (*Musa acuminate*) and common banana (*Musa paradisiac*). The high-interpolar methanolic extract from fruit peels of red banana *Musa acuminate* was revealed better proficiency antiskin-disease fungi activity as compared with common banana (*Musa paradisiac*) as shown in Figure 12.1. Effective antifungal activity was detected with *C. albicans*, *T. rubrum*, *M. gypseum*, *T.mentagrophytes*, and *T. tonsurans*. Figure 12.2 indicates bacterial activity of *E. Coli*, *B. subtilis*, *P. aeruginosa*, and *S. aureus*. Antimicrobial action was recognized to be relatively dependent.

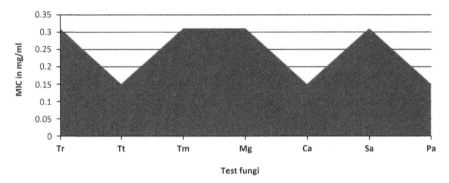

FIGURE 12.1   MIC, high-interpolar methanolic extract from peels of *Musa acuminate* fruits.

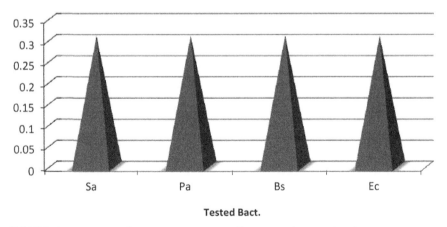

FIGURE 12.2   MBC, high-interpolar methanolic extract from peels of *Musa acuminate* fruits.

To avoid errors in the revision, the pessimistic rule (DMF) was not actively noticed. Fighting fit notorious antifungal (Ketoconazole) and antibacterial (streptomycin) prescription confirmed next to the 30 mm in diameter action.

The results on MIC and minimum inhibitory fungicide concentration, and minimum inhibitory bactericidal concentration are shown in Figures 12.1 and 12.2.

Universally, antimicrobial commotion divulges suitable to the incidence of the deliberation echelon of the secondary metabolites. Likewise, interpolar exudates indicated presence of alkaloids, flavonoids, phenols, triterpenoids, steroids, and also saponins commencing. The current results also verify the presence of secondary metabolites.[17]

In previous studies, large number of antibacterial activity has been detected in fruit peels in banana.[18] Whereas in this chapter, the first antidermatophytic activity had been conducted using high-interpolar methanolic extracts from fruit peel extracts of two varieties of banana. Among two varieties, results were effective in *Musa acumunata* or red fruit peels.

The effective herbicidal constituents of Musa spp were screened against pathogenic microbes. The phytoconstituents in the fruit peel extract showed positive response to the phenols, triterpenoids, and flavonoids concentrations and indicated foremost function during the antidermatophytic route.

12.4   SUMMARY

This chapter focuses on the evaluation of in-vitro antidermatophytic activity of high inter-polar methanolic peel extracts of red banana and common banana. The 98% methanolic soxhlet extract of peels of red banana (*Musa acuminate*) and common banana varieties (*Musa paradisica*) was appraised for skin diseases caused by fungi and bacteria. Utmost action was pragmatic in red banana crude. The MICs, minimum fungicidal concentrations, and minimum bactericidal concentrations were indomitable aligned with chosen assessment microbes. The current study affords the source intended for the seclusion and refinement of antidermatophytes active molecules from the fruit peels of red banana varieties.

ACKNOWLEDGMENT

Authors are thankful to Vice Chancellor, Registrar of Palamuru University and Gulbarga University, India for the financial support.

KEYWORDS

- **antimycotic activity**
- **high interpolar methanolic**
- **MBC**
- **MFC**
- **MIC**
- *Musa paradisica*

REFERENCES

1. Calzada, F.; Yepez-Mulia, L. Effect of Mexican Therapeutic Plant Used to Treat Trichomoniasis on *Trichomonas Vaginalis* Trophozoites. *J. Ethnopharm.* **2007,** *113,* 248–251.
2. Farnsworth, N. Screening Plant for New Medicines. In *Biodiversity*; Wilson, E. O., Ed.; Natural Academy Press: Washington D.C., 1998; pp 83–97.
3. Grabue, G. E. Treatment of Oral *Candida Mucositis* Infections. *Drugs* **1994,** *47,* 734–740.
4. Horborne, J. B. *Phytochemical Methods: Guide to Modern Techniques of Plant Analysis,* 3rd ed.; Chapman and Hall: London, 1998; p 216.
5. Karadi, R. V.; Shah, A.; Parekh, P. Antimicrobial Activities of *Musa Paradisiaca* and *Cocos Nucifera. Int. J. Res. Pharmaceut. Biomed. Sci.* **2011,** *2,* 264–267.
6. Karuppiah, P.; Mustaffa, M. Antibacterial and Antioxidant Activities of *Musa* sp. Leaf Extracts against Multidrug Resistant Clinical Pathogens Causing Nosocomial Infection. *A. Pac. J. Trop. Biomed.* **2013,** *3,* 737–742.
7. Khare, C. P. *Indian Medicinal Plants.* Springer Science, Business Media: New York, 2007; p 426.
8. Krishnan, K.; Vijayalakshmi, N. R. Alterations in Lipids and Lipid Peroxidation in Rats Fed with Flavonoids Rich Fraction of Banana (*Musa Paradisiaca*) from High Background Radiation Area. *Ind. J. Med. Res.* **2005,** *122,* 540–546.
9. Kubmarawa, D.; Ajoku, G. A.; Enwerem, N. M.; Okorie, D. A. Preliminary Phytochemical and Antimicrobial Screening of 50 Therapeutic Plants from Nigeria. *Afric. J. Biotechnol.* **2007,** *6,* 1690–1696.
10. Lavanya, K.; Beaulah, G.; Vani, G. *Musa Paradisiaca*: Review on Phytochemistry and Pharmacology. *W. J. Pharmaceut. Med. Res.* **2016,** *2* (6), 163–173.
11. Magaji, M. G.; Anuka, J. A.; Abdu-Aguye, I.; Yaro. A. H.; Hussaini, I. M. Preliminary Studies on Anti-Inflammatory and Analgesic Activities of *Securinega Virosa* in Experimental Animal Models. *J. Therapeut. Plants Res.* **2008,** *2* (2), 39–44.
12. Magaldi, S.; Mata-Essayag, S.; Hartung de Capriles, C.; Perez, C.; Colella, M. T.; Olaizola, C.; Ontiveros, Yudith. Well Diffusion for Antifungal Susceptibility Testing. *Int. J. Infect. Dis.* **2004,** *8,* 39–45.

13. National Committee for Clinical Laboratory Standards (NCCLS). *Approved Standard M2-A6*, 5th ed.; Wayne, P. A; National Committee for Clinical Laboratory Standards; 1997; p 13.

14. Padam, B. S.; Tin, H. S.; Chye, F. Y.; Abdullah, M. I. Antibacterial and Antioxidative Activities of the Various Solvent Extracts of Banana (*Musa Paradisiaca cv.* Mysore) Inflorescences. *J. Biol. Sci.* **2012,** *12* (2), 62–73.

15. Rippon, J. W. Dermatophytosis and Dermatomycosis. In *Medical Mycoses: The Pathogenic Fungi and the Pathogenic Actinomycetes*; Rippon, J. W., Ed.; WB Saunders Company: Philadelphia, PA, 1982; pp 140–248.

16. Seetharam, Y. N.; Kotresh, K.; Upalaonkar, S. B. *Flora of Gulbarga District*. Gulbarga University Press: Gulbarga, India, 2000; p 89.

17. Subrata, K. B.; Chowdhury, A. Investigation of Antibacterial Activities of Ethanol Extracts of *Musa Paradisiaca* Lam. *J. Appl. Pharmaceut. Sci.* **2011,** *1* (6), 133–135.

18. Umamaheswari, A. Phytochemical Screening and Antimicrobial Effects of *Musa Acuminata. Int. Res. J. Pharm.* **2017,** *8* (8), 41–44.

19. Yusuf, M.; Begum, J.; Hoque, M. N.; Chowdhury, J. U. *Medicinal Plants of Bangladesh*. 2nd ed.; BCSIR Laboratories: Dhaka, Bangladesh, 2009; pp 462–463.

CHAPTER 13

SECONDARY METABOLITES FROM LICHEN GENUS (*RAMALINA* ACH.): APPLICATIONS AND BIOLOGICAL ACTIVITIES

T. R. PRASHITH-KEKUDA and K. S. VINAYAKA

ABSTRACT

Lichens are composite organisms comprising of mycobiont (a fungal partner) and a photobiont (an algal or cyanobacterial partner) and represent an ecologically stable and obligate symbiotic interaction. Lichens produce characteristic lichen metabolites that do not occur normally in other organisms. The lichen genus ((*Ramalina* Ach.): Fungi (unicellular organisms) > Ascomycetes (nonmotile spores) > Lecanorales (fungi forming) > Ramalinaceae) is widely studied lichen genera comprising of >240 species. Primary and secondary metabolites were identified, such as salazinic acid, usnic acid, sekikaic acid, etc. Several *Ramalina* species are used traditionally as medicine, food, and spice. Biological activities of *Ramalina* Ach. have been reported, such as anti-inflammatory, antiviral, antimicrobial, anti-oxidant, enzyme inhibitory, insecticidal, cytotoxic, anthelmintic, and immunostimulatory. The present chapter has focused on the chemistry, ethnobotanical uses, and pharmacological activities of various *Ramalina* species.

13.1 INTRODUCTION

Lichens are self-supporting, ecologically obligate, outstandingly successful, and stable symbiotic association of a photosynthetic partner (a photobiont: a cyanobacterium or an alga) and a fungal partner (a mycobiont: an ascomycete or basidiomycete member). It is estimated that about 19% of all fungi have been lichenized. Lichens are cosmopolitan in distribution and occur in tropical and temperate regions, deserts, arctic regions, and high mountains. Together with mosses, lichens cover about 8–10% of terrestrial ecosystems. Lichens usually occur in one of the three types of growth forms: viz. crustose, foliose, and fruticose (Fig. 13.1; Appendix I) and are capable of growing on rocks (saxicolous), soil (terricolous), barks (corticolous), and leaves (foliicolous). Lichens are considered valuable as bioindicators of air pollution. Lichens are used traditionally for purposes, such as food, fodder, spices and flavoring agents, sources of dyes, for decorative purposes, preparation of perfumes and alcohol, and as medicine to treat several diseases or disorders.

| Crustose | Foliose | Fruticose |

FIGURE 13.1   Growth forms of lichens.
Source: Photographs by Prashith-Kekuda.

Lichens have been used in traditional medicine in several countries. Lichens can synthesize wide range of primary and secondary metabolites. Many of these metabolites are characteristic to lichens (often referred as lichen substances) and do not occur in other organisms. More than 1000 of such metabolites have been identified in lichens (Appendix II, III). Crude solvent extracts and isolated compounds[8,21,30,45,52,62,68,70,75,77,93,94,95,102,116] from various lichen species were shown to display several biological activities, such as insecticidal, antimicrobial, anthelmintic, antiviral, antitumor, enzyme inhibitory, antiherbivore, and immunostimulatory. *Ramalina* species and the common names are indicated in Appendix III.

This chapter focuses on bioactivities of secondary metabolites from Lichen genus (*Ramalina* Ach.).

13.2 THE GENUS *RAMALINA* ACH.

The lichen genus *Ramalina* Ach. is one of the ubiquitous and most widely studied lichen genera and was first described by Acharius. The genus

Ramalina consists of about 240 species that are distributed in various parts of the world (Appendix III). The members of *Ramalina* are characterized by a branched fruticose thallus (Fig. 13.2) that is attached to the substratum by means of a basal holdfast. The thallus is pendent or erect, heteromerous, tufted or sparsed, and dichotomously or irregularly branched. The branches may be circular, wide lobed, or narrow strap shaped. Species of *Ramalina* occur on rocks (saxicolous), peaty soil (terricolous), wood (lignicolous), and bark (corticolous). The thalli are greenish-gray to yellowish-gray in color when fresh; however, the color changes to yellowish-brown to dark-brown on drying. The species of *Ramalina* contain pseudocyphellae. Usnic acid is the major diagnostic secondary metabolite found in *Ramalina*.[3, 59, 67]

R. nervulosa   *R. hossei*

R. pacifica   *R. conduplicans*

FIGURE 13.2 *Ramalina* sp.
Source: Photographs by Vinayaka.

13.3 TRADITIONAL USES OF *RAMALINA* SPECIES

In several countries, several lichens are popularly used as flavoring agent and are incorporated in certain cuisine. Lichens are important components in *garam* masala, *sambhar masala*, and *meat masala*, which are mixtures of powdered condiments and spices.

Potential use of lichens has been described in traditional systems of medicine like Unani and Ayurveda. Ethnic communities in different parts of the world use lichens for the treatment of several ailments, such as bleeding piles, stomach disorders, bronchitis, skin diseases, wound, and jaundice.[23,29,36,37,86,89,104,109,118] Species of *Ramalina* are known to be traditionally used in North America, Asia, and Europe as sources of food, flavoring agent, and medicine.[18] as indicated in Table 13.1.

TABLE 13.1   The Traditional Uses of Some *Ramalina* Species.

Ramalina species	Region	Traditional use	Ref.
R. botrytis	Central Nepal	Vegetable.	[97]
R. bourgeana (thallus)	Eastern Andalucia of Spain	Decoction is used as diuretic for the treatment of renal lithiasis.	[36]
R. calicaris, R. commixta, R. fastigiata, R. minuscula R. sinensis	Guizhou province, China	Fodder; wild vegetable; flavoring agent.	[66]
R. conduplicans	Yunnan province of China	Food.	[117]
R. sinensis	Rai and Limbu communities of East Nepal	Edible vegetable.	[6]
R. farinacea	Podrinje in the Canyon of Drina River	Edible lichen; the form of mush and lichen bread.	[83]
R. himalayensis	Tribal communities in West Kameng district of Arunachal Pradesh, India	Edible vegetable after boiling and air drying.	[85]
R. pacifica	Central Western Ghats of Karnataka, India	Spice; flavoring agent; decorative purpose; religious activities.	[115, 116].
R. subcomplanata	Tribal communities of Madhya Pradesh, India	Spice; flavoring agent.	[109]
Ramalina conduplicans	Central Indian regions	Spice for meat and vegetables.	[56]
Ramalina conduplicans	Sikkim and Darjeeling regions of India	Medicine for jaundice and hydrophobia.	[86]
Ramalina species	Local communities in Naban river watershed national nature reserve, Yunnan, China	Wild food.	[33]
	Ethnic communities from Hangdewa and Phurumbu, Taplejung district, Nepal	Ethnomedicine as antiseptic tincture to heal wounds.	[23]

13.4 CHEMISTRY OF *RAMALINA* SPECIES

Lichens produce primary and secondary metabolites. The primary metabolites include: polysaccharides (beta-glucans, alpha-glucans, and galactomannans), amino acids, vitamins, proteins, and carotenoids. Most of the secondary metabolites in lichens are small molecules and are unique to lichens (i.e., formed only in lichenized state) and do not occur in other organisms. Metabolic pathways (viz., acetyl-polymalonyl, mavalonic acid, and shikimic acid pathways) are involved in the formation of lichen substances. These lichen metabolites have low molecular weight and may be found in cortex or medulla. The presence and type of these secondary metabolites is also useful in the identification and taxonomy of lichens. Chromatographic (such as TLC, column chromatography, and HPLC) and spectral techniques (such as H-NMR, C-NMR, and MS) have been employed for isolation, identification, and elucidation of structures of lichen metabolites.[5,49,62,69,70,87,106]

Species of *Ramalina* are capable of producing a variety of secondary metabolites, belonging to three major classes, namely: depsides, depsidones, and dibenzofurane. Usnic acid, a dibenzofurane, is the signature lichen substance found in almost all *Ramalina* species. Besides usnic acid, compounds, such as sekikaic acid, atranorin, and salazinic acid are among the commonly found metabolites in some *Ramalina* species. Some compounds are unique to certain *Ramalina* species.[3,11,63,72,101,106] Chester and Elix[15] revealed the presence of 4,4'-di-O- methylcryptochlorophaeic acid, 2-O-methylsekikaic acid, 2,4'-di-O-methylnorsekikaic acid, and 4'-O-methyl-paludosic acid in several *Ramalina* species. Biocompounds detected in some *Ramalina* species are listed in Table 13.2.

TABLE 13.2 Biocompounds Detected in Some *Ramalina* Species.

Species	Compounds	Ref.
R. Africana	Usnic acid, divaric acid, ethyl divaricatinate	[96]
R. Americana	Trivaric acid	[20]
R. capitata	Evernic acid, orcinol, orcinol monomethyl ether, 3-methylorsellinic acid, usnic acid	[11, 119]
R. celastri *R. fastigiata* *R. fraxinea* *R. mahoneyi* *R. menziesii* *R. pollinaria* *R. polymorpha* *R. sinesis*	Usnic acid	[11, 28, 81, 92, 93]

TABLE 13.2 *(Continued)*

Species	Compounds	Ref.
R. chilensis	Usnic acid, ramalinolic acid	[87]
R. conduplicans	Usnic acid, atranorin, barbatic acid, evernic acid, diffractaic acid, sekikaic acid, salazinic acid	[46, 106]
R. dumeticola	Atronorin, hyperhomosekikaic acid, sekikaic acid	[39]
R. farinacea	Usnic acid, protocetraric acid, norstictic acid	[107]
R. fastigiata	Usnic acid, methyl evernate, obtusatic acid	[84]
R. fraxinea	Usnic acid, sekikaic acid, protocetraric acid, atranorin, evernic acid, obtusatic acid	[84]
R. glaucescens	5-chlorosekikaic acid, usnic acid, sekikaic acid, atranorin	[24]
R. hierrensis	Ursolic acid, usnic acid, divaricatic acid, atranorin, hierridin, stictic acid, gangaleodin, variolaric acid, montagnetol, parietin	[35]
R. hossei R. hyrcana	Usnic acid, sekikaic acid	[53, 110]
R. inflate	Salazinic acid, divaricatic acid, some terpenes	[31]
R. javanica	Vicanicine	[103]
R. lacera	Bourgeanique acid, diffractaic acid, divaricatic acid, lecanoric acid, usnic acid, norstictic acid, protocetric acid	[41, 105]
R. leiodea	Boninic acid, usnic acid, 2-0-methylsekikaic acid, 2,4′-di-0-methylnorsekikaic acid, 4′-0-methylpaludosic acid, 4,4′-di-0-methylcryptochlorophaic acid	[7]
R. luciae	Sekikaic acid, 4′-O-demethylsekikaic acid, 4′-O-methylnorsekikaic acid, usnic acid, ramalinolic acid	[7]
R. nervulosa, R. pacifica	Usnic acid, sekikaic acid	[114]
R. pollinaria	Protocetraric acid, norstictic acid, salazinic acid, evernic acid	[110]
R. roesleri	Atranorin, protolichesterinic acid, usnic acid, homosekikaic acid, sekikaic acid	[34, 98]
R. siliquosa	Usnic acid, atranorin, arabitol, peristictic acid, chloroatranorin, conhypoprotocetraric acid, salazinic acid, stictic acid, cryptostictic acid, protocetraric acid, hypoprotocetraric acid, 4-O-demethylbarbatic acid, norstictic acid	[19, 76]
R. sorediosa	Usnic acid, homosekikaic acid, salazinic acid	[27]
R. terebrata	Parietin, usnic acid, atraric acid, inositol	[17]
R. usnea	Usnic acid, 2-Hydroxy-4-methoxy-6-propyl-methyl benzoate	[64]

13.5 PHYTOCHEMICAL ACTIVITIES OF *RAMALINA* SPECIES

Crude extracts and purified compounds from several *Ramalina* species have shown experimentally to display wide range of bioactivities, such as antimicrobial, anti-oxidant, antiviral, antischistosomal, insecticidal, anthelmintic, anti-inflammatory, immunostimulatory, enzyme inhibitory, and cytotoxic activities. Some of the bioactivities of various *Ramalina* species are described in this section.

13.5.1   ANTIBACTERIAL ACTIVITY OF RAMALINA SPECIES

Extracts and metabolites from *Ramalina* species possess antibiotic properties even against drug-resistant pathogens. The antibacterial potential of *Ramalina* species has been investigated by various researchers using techniques such as disc diffusion, agar well diffusion, and broth microdilution.

RF1 and RF2, two bioactive column chromatographic fractions of *R. farinacea*, were shown to inhibit clinical isolates of *Staphylococcus aureus*.[25] Acetone extract of *R. farinacea* was effective in inhibiting certain bacteria.[107] Extracts of *R. sorediosa* were effective against *Bacillus subtilis* and *S. aureus*.[27] Extracts obtained from *R. pollinaria* and *R. polymorpha* were shown to inhibit some bacteria.[38]

Cansaran et al.[11] evaluated antibacterial activity of acetone extract of *R. capitata*, *R. pollinaria*, *R. polymorpha*, *R. fraxinea*, and *R. fastigiata* against Gram-positive and Gram-negative bacteria. All lichens were effective in inhibiting at least one of the seven test bacteria. *R. fastigiata* showed highest inhibitory effects to test bacteria. The study also revealed a direct correlation between antibacterial activity and usnic acid content of lichens. Extract of an Antarctic lichen *R. terebrata* was effective in inhibiting gram-positive bacteria (*S. aureus* and *B. subtilis*).[80]

Karagöz et al.[48] showed antibacterial activity of ethanol and aqueous extracts of *R. farinacea* against *B. subtilis* and *S. aureus*. The study carried out by Cobanoglu et al.[16] revealed inhibitory activity of acetone extract of *R. farinacea* against Gram-negative bacteria. The study carried out by Hoskeri et al.[44] showed the potential of petroleum ether and ethanol extract of *R. pacifica* to inhibit reference strains and clinical strains of bacteria. Shrestha and St. Clair[93] revealed marked inhibitory activity of acetone extract of *R. menziesii* against *P. aeruginosa*, *S. aureus*, and MRSA while the extract was not much effective against *Escherichia coli*. The hexane extract of *R. roesleri* was shown to be highly active against *Staphylococcusaureus* and *Streptococcus mutans*.[98] In a study, acetone extract of *R. sinensis* was effective in inhibiting *Pseudomonasaeruginosa, Staphylococcusaureus*, and methicillin resistant *S. aureus*.[92] Methanol extract of *R. hossei* was effective in inhibiting *Streptococcusmutans* (dental caries), *Staphylococcusaureus* (burn), and urinary tract pathogens.[47]

Ristić et al.[84] showed the antibacterial activity of extract of *R. fastigiata* and *R. fraxinea* against a panel of bacteria. Ganesan et al.[31] showed the potential of petroleum ether, methanol, ethyl acetate, 2-propanol, and benzene extracts of *R. inflate* to inhibit bacteria by agar well diffusion method. Sesal et al.[88] observed the antibacterial activity of chloroform and methanol extracts of *R. canariensis, R. chondrina, R. fastigiata*, and *R. fraxinea* to inhibit at least one of the tested bacteria. Kambar et al.[46] and Devi et al.[22] showed activity of various solvent extracts obtained from *R. conduplicans* to inhibit bacteria. Ethanol extracts of *R. hossei, R. conduplicans*, and *R. pacifica* exhibited inhibitory activity against *Streptococcus mutans* isolates.[50]

Acetone extract of *R. dumeticola* was effective against gram-positive bacteria.[39] Ankith et al.[1] showed antibacterial potential of methanol extract obtained from *R. hossei, R. conduplicans*, and *R. pacifica* against bacteria. The acetone extract of *R. farinacea* showed inhibitory activity against *Streptococcus pneumoniae* and MRSA.[108] Acetone extract of *R. capitata* was effective against gram-positive bacteria but not for gram-negative bacteria.[120] Hengameh and Rajkumar[42] observed marked inhibition of bacteria by extracts of *R. celastri*. Isolated constituents from *Ramalina* species have also shown to exhibit antibacterial activity. Table 13.3 shows antibacterial activity of some biocompounds isolated from *Ramalina* species.s

TABLE 13.3   Antibacterial Potential of Metabolites of Some *Ramalina* Species.

Species	Compound	Activity against	Ref.
R. farinacea	Norstictic acid, usnic acid	Gram-positive and Gram-negative bacteria	[107]
R. fastigiata	Obtusatic acid, methyl evernate		[84]
R. glaucescens	Usnic acid, sekikaic acid	*Bacillus subtilis*	[24]
R. terebrata	Ramalin, usnic acid	*Bacillus subtilis*	[79]

13.5.2   ANTIFUNGAL ACTIVITY OF RAMALINA SPECIES

Several methods such as disc diffusion, well diffusion, and poisoned food technique have been employed to evaluate antifungal activity of extracts and purified compounds from *Ramalina* species. It is evident from many studies that *Ramalina* species exhibit antifungal activity against human and plant pathogens and storage fungi. Gulluce et al.[38] showed that extracts obtained from *R. pollinaria* and *R. polymorpha* inhibit some yeast and fungi. Various extracts of *R. hossei* exhibited inhibitory against *Aspergillus niger* and *A. fumigatus*. Among fungi, *A. niger* was inhibited to highest extent.[54] *R. conduplicans* inhibited the growth of *Colletotrichum capsici*, *Helminthosporium* sp., *Alternaria* sp., *Aspergillus flavus*, and *Sclerotium rolfsii*.[46] *R. hossei* displayed antifungal activity toward chilli anthracnose pathogen *Colletrotrichum capsica*.[47] Acetone extract of *R. conduplicans* was effective against *Trichophyton mentagrophytes* and *T. rubrum*.[77]

Devi et al.[22] showed antifungal potential of *R. conduplicans* against some plant pathogens. Chloroform and methanol extracts of *R. canariensis* and *R. chondrina* exhibited inhibitory activity against *Candida albicans*[88] Ganesan et al.[31] showed antifungal potential of *R. inflata* against yeasts and molds. Ethyl acetate extract of *R. farinacea* caused inhibition of two phytopathogenic fungi, viz., *F. oxysporum* and *F. solani*.[90] The study carried out by Ristić et al.[84] found the antifungal activity of *R. fastigiata* and *R. fraxinea* against *C. albicans* and a panel of molds. Methanol extracts of *R. hossei*, *R. conduplicans*, and *R. pacifica* were shown to inhibit the mycelial growth of seed-borne fungi, viz., *Alternaria* sp., *Curvularia* sp., and *Fusarium* sp.[1] Table 13.4 indicates antifungal potential of some biocompounds isolated from *Ramalina* species.

TABLE 13.4   Antifungal Activity of Metabolites of Some *Ramalina* Species.

Species	Compound	Bioactivity against	Ref.
R. farinacea	Protocetraric acid	*C. albicans* and *C. glabrata*	[107]
R. fastigiata	Obtusatic acid, methyl evernate	*C. albicans* and some molds	[84]
R. roesleri	Usnic acid	*Fusarium udum, Sclerotium rolfsii, Pythium aphanidermatum, Rhizoctonia solani*	[34]
	Sekikaic acid	*S. rolfsii, F. udum*	
	Homosekikaic acid	*S. rolfsii, Rhizoctonia solani*	
R. glaucescens	Usnic acid	*T. mentagrophytes* and *Cladosporium resinae*	[24]

13.5.3   ANTIVIRAL ACTIVITY OF RAMALINA SPECIES

Usnic acid isolated from *R. celastri* showed antiviral activity against viruses Junin and Tacaribe.[28] Ethylacetate-soluble fraction from *R. farinacea* was tested for antiviral activity against HSV-1 and RSV by Esimone et al.[26] The fraction was effective in inhibiting HSV-1 and RSV. The fraction was found to target entry as well as postentry steps of the replication cycle in HIV-1 and the entry step of replication cycle in the RSV. Two sub-fractions (viz., rfO and rfM) of ethyl acetate fraction were active against lentiviral vector and HIV-1 and HSV-1, respectively. In another study, Lai et al.[55] investigated antiviral activity of phenolic compounds from *R. farinacea* against RSV. Sekikaic acid caused potent inhibition of a recombinant strain of RSV and RSV A2 strain. Sekikaic acid was found to interfere with viral postentry step of viral replication.

13.5.4 ANTI-OXIDANT ACTIVITY OF RAMALINA SPECIES

Methanol extracts of *R. conduplicans* and *R. hossei* exhibited anti-oxidant activity as revealed by DPPH assay and ferric reducing assay.[53] Extract obtained from *R. conduplicans* exhibited potent radical scavenging activity, antilinoleic acid peroxidation activity, and reducing potential.[61] Methanol and acetone extracts from *R. peruviana* revealed effective scavenging of DPPH radicals and bleaching of beta-careotene.[100]

Extracts of *R. celastri*, *R. stenospora* and *R. americana* were effective in scavenging ABTS radicals and inhibiting lipid oxidation.[111] Based on in-vitro assays, Verma et al.[112] revealed anti-oxidant activity of solvent extracts of *R. celastri*, *R. nervulosa*, and *R. pacifica*. Acetone extract of *R. fastigiata* and *R. fraxinia* displayed scavenging of DPPH radicals with an IC$_{50}$ value of 423.51 and 285.45 µg/mL, respectively. The extracts were also shown to exhibit reducing power.[84]

Smitha and Garampalli[99] revealed DPPH scavenging and ferric-reducing potential of *R. pacifica*. Extracts of *R. inflata* exhibited anti-oxidant activity in some in-vitro assays.[31] Gunasekaran et al.[39] showed scavenging of DPPH radicals by acetone and methanol extracts of *R. dumeticola*.

Ethanol, methanol, and acetone extracts of *R. subpusilla* and *R. conduplicans* exhibited scavenging potential against DPPH and superoxide radicals.[82] Acetone extract of *R. capitate* showed radical scavenging and anti-oxidant activity.[120] Anti-oxidant and free radical scavenging activity of pure compounds from *Ramalina* species was investigated by researchers and it was observed that the compounds possess promising activity (Table 13.5).

TABLE 13.5 Radical Scavenging and Anti-Oxidant Activity Displayed by *Ramalina* Metabolites.

Species	Extract/metabolite	Ref.
R. celastri	Usnic acid	[112]
R. conduplicans	Sekikaic acid, homosekikaic acid	[61]
	Usnic acid, salazinic acid, atranorin, diffractaic acid, barbatic acid, evernic acid	[106]
R. dumeticola	Atronorin, sekikaic acid, hyperhomosekikaic acid	[39]
R. fastigiata	Methylevernate, obtusatic acid	[84]
R. nervulosa	Sekikaic acid	[112]
R. pacifica		
R. roesleri	Sekikaic acid, homosekikaic acid	[98]
R. terebrata	Ramalin	[78]

13.5.5 ENZYME INHIBITORY ACTIVITY OF RAMALINA SPECIES

Studies have shown the potential of extracts and biocompounds from *Ramalina* species to inhibit several enzymes such as amylase, lipase, tyrosinase, and cholinesterase of clinical significance. Verma et al.[112] revealed the potential of acetone, ethanol, and methanol extracts of *R. celastri*, *R. nervulosa*, and *R. pacifica* to inhibit α- and β-glucosidases. Vinayaka et al.[113] showed inhibition of amylase activity by methanol extract of *R. hossei* and *R. conduplicans*. Methanol extracts of *R. hyrcana* and *R. pollinaria* were effective in inhibiting amylase activity in a concentration dependent manner.[110] Hexane, ethyl acetate, ethanol, and methanol extracts of *R. sinensis* were effective in inhibiting α-amylase in a dose-dependent manner.[43]

Parizadeh and Garampalli[71] revealed the potential of solvent extracts of *R. sinensis* to inhibit the activity of β-glucosidase. Ethyl acetate and methanol extracts of *R. celastri* were screened for inhibitory activity against pancreatic lipase. Both extracts inhibited enzyme dose dependently with marked activity shown by methanol extract.[91] Ramalin isolated from the Antarctic lichen *R. terebrata* was shown to have tyrosinase inhibitory activity, cell-free tyrosinase activity, and intracellular tyrosinase activity. At a noncytotoxic concentration, ramalin resulted in a decrease in melanin synthesis in melan-a-cells in a dose-dependent manner. Ramalin also possesses skin brightness property.[14] Zrnzevic et al.[120] showed that extract of *R. capitata* inhibited the activity of pooled human serum cholinesterase.

13.5.6   CYTOTOXIC ACTIVITY OF RAMALINA SPECIES

It is shown that crude extracts and purified chemicals from several *Ramalina* species possess cytotoxic properties. Solvent extracts of *R. cuspidata* were shown to be effective against murine and human cancer cell lines as revealed by MTT assay.[4] Crude extract from *R. glaucescens* was shown to exhibit moderate cytotoxicity against P388 murine leukemia cell line.[24] Karagöz et al.[48] showed cytotoxicity of aqueous and ethanol extracts of *R. farinacea* against vero cells. Ethanol extract displayed stronger cytotoxicity. Methanol extract of *R. farinacea* was toxic to mammalian cell lines and mouse splenocytes.[66] Solvent extract of *R. fastigiata* and R. *fraxinea* displayed cytotoxicity against cell lines, viz., HeLa, A549, and LS174 in MTT assay.[84] Table 13.6 shows the cytotoxic effect of purified biocompounds from *Ramalina* species.

TABLE 13.6   Cytotoxic Activity of *Ramalina* Metabolites.

Species	Biocompound	Cytotoxicity against	Ref.
R. celastri	α-D-glucan and its sulphated derivative	HeLa cells	[57]
R. farinacea	Usnic acid	A549 and V79 cells	[51]
	Obtusetic acid and methyl evernate	HeLa, A549, LS174 cells	[84]
R. terebrata	Ramalin	HCT116 cells	[102]
R.glaucescens	Usnic acid, atranorin, sekikaic acid, 5-chlorosekikaic	P388 murine leukemia cell line	[24]
Ramalina sp.	Norstictic acid	UACC-62 cells	[9]

13.5.7   ANTISCHISTOSOMAL ACTIVITY OF RAMALINA SPECIES

A polysaccharide designated as α-D-glucan was recovered from *R. celastri* and investigated for antischistosomal activity, in terms of egg elimination, worm burden, and hepatic granuloma formation in *Schistosoma mansoni*-infected mice. It was observed that the treatment with both free as well as liposome-encapsulated sulphated polysaccharide caused reduction in the number of granulomas and the result was statistically significant.[2]

13.5.8   INSECTICIDAL ACTIVITY OF RAMALINA SPECIES

Dichloromethane and methanol extracts of *Ramalina* species exhibited larvicidal activity against larvae (II instar) of *Aedes aegypti*[65] Crude methanolic extracts of *R. pacifica* and *R. nervulosa* were effective in causing mortality of *A. aegypti* larvae.[114] Usnic acid isolated from *R. farinacea* has shown potent activity against III-IV larval stages of *Culex* pipiens.[13] Methanol extracts from *R. hossei* and *R. conduplicans* were shown to cause dose-dependent mortality of second instar larvae of *A. aegypti*.[52] Usnic acid and 2-Hydroxy-4-methoxy-6-propyl-methyl benzoate, isolated from *R. usnea* were effective larvicides against third instar larvae of *A. aegypti*[64]

13.5.9   ANTI-INFLAMMATORY ACTIVITY OF RAMALINA SPECIES

Studies have revealed the anti-inflammatory potential of purified compounds from *Ramalina* species. Ramalin was isolated from the Antarctic lichen *R. terebrata*is and screened for in-svivo and in-vitro anti-inflammatory activities against atopic dermatitis in mice. On administering Ramalin orally, a reduction in scratching behavior and reduction in IL-4 and serum immunoglobulin E levels and mRNA levels of IL-10 and IL-4 was observed. Treatment with Ramalin resulted in significantly less cytokines and inflammatory chemokines under in-vitro conditions. Ramalin also inhibited the activation of nuclear factor-kappa B and the phosphorylation of MAPK. It is inferred that the ramalin seems to modulate the production of immune mediators in terms of inhibition of the nuclear factor-kappa B and MAPK signaling pathways.[74] Stereocalpin A, an antitumor compound recovered from *R. terebrata*, inhibited the expression of adhesion molecules (namely VCAM-1 and intercellular adhesion molecule-1) induced by TNF-α in activated vascular smooth muscle cells. It is inferred that stereocalpin A was protective by modifying inflammation within the atherosclerotic lesion.[10] The compound Ramalin was also shown to be effective in inhibiting VCAM-1 expression and adhesion of monocyte to vascular smooth muscle cells, which suggests that the compound ramalin could be a potential agent for modulation of inflammation within atherosclerosis.[73]

13.5.10   GASTROPROTECTIVE ACTIVITY OF RAMALINA SPECIES

Various solvent extracts of *R. capitata* were investigated for gastroprotective activity in rats. Ethanol—water extract, @ 20 mg/kg dose, exhibited significant gastroprotective activity to cause reduction in gastric lesions (66.6%) induced by indomethacin in rats and the protective effect was linked to neutrophil infiltration into gastric tissues as well as its protective effect on oxidative damage.[40]

13.5.11   ANTIGENOTOXIC ACTIVITY OF RAMALINA SPECIES

The extract obtained from R. capitata displayed antigenotoxic effect against colloidal bismuth subcitrate induced genotoxicity.[32] Zrnzevic et al.[120] also highlighted the protective property of acetone extract of R. capitata for reduction in micronucleus frequency.

13.5.12   IMMUNOSTIMULATORY ACTIVITY OF RAMALINA SPECIES

Carlos et al.[12] revealed immunostimulatory potential of norstictic acid from Ramalina sp. in murine macrophage cells. Treatment of cells with this compound showed an induction of production of nitric oxide and hydrogen peroxide both of which are involved in immune functions.

13.5.13   ANTHELMINTIC ACTIVITY OF RAMALINA SPECIES

Vinayaka et al.[114] showed dose-dependent anthelmintic activity of methanol extract of R. pacifica and R. nervulosa as indicated by paralysis and death of adult earthworms. In another study, Kumar et al.[53] observed dose-dependent anthelmintic activity of R. conduplicans and R. hossei against adult Indian earth worms.

13.5.14   DERMAL FIBROBLAST PROLIFERATION ACTIVITY OF RAMALINA SPECIES

Usimine-C, isolated from R. terebrata, was able to induce cell proliferation of human dermal fibroblast, CCD-986SK. Treatment with the compound resulted in an increase in synthesis of type-I procollagen indicating that the compound might have triggered an internal signal for type-I procollagen expression.[58]

13.5.15   *TAU PROTEIN INHIBITION BY RAMALINA SPECIES*

Parietin was isolated from *R. terebrata* and was screened for *tau* protein (a protein involved in Alzheimer's disease) aggregation inhibitory activity. Parietin was able to cause inhibition of aggregation process of tau in a concentration range 3 µg/mL to 28 µg/mL.[17]

13.6   SUMMARY

Extensive literature survey on lichens revealed the presence of diverse type of chemical compounds in *Ramalina* species. Some species of *Ramalina* are traditionally used as food, flavoring agents, and as medicine to treat diseases or disorders. It is evident from several studies that the species of the genus *Ramalina* exhibited several pharmacological activities including cytotoxic activity. The presence of bioactive lichen substances accounts for the observed bioactivities of *Ramalina* species.

KEYWORDS

- lichens
- antibacterial
- antifungal
- anti-oxidant
- salazinic acid
- sekikaic acid
- usnic acid

REFERENCES

1. Ankith, G. N.; Rajesh, M. R.; Karthik, K. N.; Avinash, H. C.; Kekuda, P. T. R.; Vinayaka, K. S. Antibacterial and Antifungal Activity of Three *Ramalina* Species. *J. Drug Deliv. Therapeut.* **2017,** *7* (5), 27–32.
2. Araújo, R. V.; Melo-Júnior, M. R.; Beltrão, E. I.; Mello, L. A. Evaluation of the Antischistosomal Activity of Sulfated α-D-Glucan from the Lichen *Ramalina Celastri* Free and Encapsulated into Liposomes. *Brazil. J. Med. Biol. Res.* **2011,** *44* (4), 311–318.

3. Awasthi, D. D. *A Compendium of the Macrolichens from India, Nepal and Sri Lanka.* Publisher Bishen Singh Mahendra Pal Singh: Dehra Dun, India, 2007; p 215.

4. Bézivin, C.; Tomasi, S.; Dévéhat, L. F.; Boustie, J. Cytotoxic Activity of Some Lichen Extracts on Murine and Human Cancer Cell Lines. *Phytomedicine* **2003,** *10* (6–7), 499–503.

5. Bhattacharyya, S.; Deep, P. R.; Singh, S.; Nayak, B. Lichen Secondary Metabolites and Its Biological Activities. *Am. J. Pharmtech. Res.* **2016,** *6* (6), 28–44.

6. Bhattarai, T.; Subba, D.; Subba, R. Nutritional Value of Some Edible Lichens of East Nepal. *Angewandte Botanik* **1999,** *73* (1), 11–18.

7. Blanchon, D. J.; de Lange, P. J.; Galloway, D. J. New Records of *Ramalina* for Mainland New Zealand. *NZ J. Bot.* **2015,** *53* (4), 192–201.

8. Boustie, J.; Grube, M. Lichens—A Promising Source of Bioactive Secondary Metabolites. *Plant Genet. Res.* **2005,** *3* (2), 273–287.

9. Brandão, L. F. G.; Alcantara, G. B.; Matos, M. F. C.; Bogo, D. Cytotoxic Evaluation of Phenolic Compounds from Lichens against Melanoma Cells. *Chem. Pharmaceut. Bull.* **2013,** *61* (2), 176–183.

10. Byeon, H. E.; Park, B. K.; Yim, J. H.; Lee, H. K.; Moon, E. Y. Stereocalpin—A Inhibits the Expression of Adhesion Molecules in Activated Vascular Smooth Muscle Cells. *Int. Immunopharmacol.* **2012,** *12* (2), 315–325.

11. Cansaran, D.; Atakol, O.; Halici, M. G.; Aksoy, A. HPLC Analysis of Usnic Acid in Some *Ramalina* Species from Anatolia and Investigation of Their Antimicrobial Activities. *Pharmaceut. Biol.* **2007,** *45* (1), 77–81.

12. Carlos, I. Z.; Carli, C. B. A.; Maia, D. C. G.; Benzatti, F. P. Immunostimulatory Effects of the Phenolic Compounds from Lichens on Nitric Oxide and Hydrogen Peroxide Production. *Brazil. J. Pharmacog.* **2009,** *19* (4), 847–852.

13. Cetin, H.; Tufan-Cetin, O.; Turk, A. O.; Tay, T.; Candan, M. Activity of Major Lichen Compounds, (−)- and (+)-Usnic Acid, Against the Larvae of House Mosquito, *Culex pipiens* L. *Parasitol. Res.* **2008,** *102,* 1277–1279.

14. Chang, Y.; Ryu, J.; Lee, S.; Park, S. G.; Bhattarai, H. D.; Yim, J. H.; Jin, M. H. Inhibition of Melanogenesis by *Ramalin* from the Antarctic Lichen *Ramalina Terebrata*. *J. Soc. Cosmet. Sci. Korea* **2012,** *38* (3), 247–254.

15. Chester, D. O.; Elix, J. A. The Identification of Four New Meta-Depsides in the Lichen *Ramalinaasahinae*. *Aus. J. Chem.* **1978,** *31* (12), 2745–2749.

16. Cobanoglu, G.; Sesal, C.; Gokmen, B.; Cakar, S. Evaluation of the Antimicrobial Properties of Some Lichens. *South West. J. Horticult. Biol. Environ.* **2010,** *1* (2), 153–158.

17. Cornejo, A.; Salgado, F.; Caballero, J.; Vargas, R. Secondary Metabolites in *Ramalina Terebrata* Detected by UHPLC/ESI/MS/MS and Identification of Parietin as Tau Protein Inhibitor. *Int. J. Mol. Sci.* **2016,** *17* (8), 1303–1310.

18. Crawford, S. D. Lichens Used in Traditional Medicine. In *Lichen Secondary Metabolites*; Rankovic, B., Ed.; Springer International Publishing: Switzerland, 2015; pp 27–80.

19. Culberson, C. F. Some Constituents of the Lichen *Ramalina Siliquosa*. *Phytochem.* **1965,** *4* (6), 951–961.

20. Culberson, C. F.; LaGreca, S.; Johnson, A.; Culberson, W. L. Trivaric Acid, A New Tridepside in the *Ramalina Americana* Chemotype Complex (Lichenized Ascomycota: Ramalinaceae). *Bryolog.* **1999,** *102* (4), 595–601.

21. Dayan, F. E.; Romagni, J. G. Lichens as a Potential Source of Pesticides. *Pestic. Outl.* **2001,** *2001,* 229–232.

22. Devi, A. B.; Mohabe, S.; Nayaka, S.; Reddy, M. A. *In Vitro* Antimicrobial Activity of Lichen *Ramalina Conduplicans* Vain. Collected from Eastern Ghats, India. *Sci. Res. Rep.* **2016,** *6* (2), 99–108.

23. Devkota, S.; Chaudhary, R. P.; Werth, S.; Scheidegger, C. Indigenous Knowledge and Use of Lichens by the Lichenophilic Communities of the Nepal—Himalaya. *J. Ethnobiol. Ethnomed.* **2017,** *13*, 15–18.

24. Dias, D. A.; Urban, S. Phytochemical Investigation of the Australian Lichens *Ramalina Glaucescens* and *Xanthoria Parietina*. *Nat. Prod. Communicat.* **2009,** *4* (7), 959–964.

25. Esimone, C. O.; Adikwu, M. U. Susceptibility of Some Clinical Isolates of *Staphylococcus Aureus* to Bioactive Column Fractions From the Lichen *Ramalina Farinacea* (L.) Ach. *Phytother. Res.* **2002,** *16* (5), 494–496.

26. Esimone, C. O.; Grunwald, T.; Nworu, C. S.; Kuate, S.; Proksch, P.; Uberla, K. Broad Spectrum Antiviral Fractions From the Lichen *Ramalina Farinacea* (L.) Ach. *Chemother.* **2009,** *55* (2), 119–126.

27. Falcão, E. P. S.; da Silva, N. H.; Gusmão, N. B.; Ribeiro, S. M.; Pereira, E. C. Antimicrobian Activity of Phenols Actives From the Liquen *Ramalina Sorediosa* (B. de Lesd.) Laundron. *Acta Botanic. Brasil.* **2004,** *18* (4), 911–918.

28. Fazio, A. T.; Adler, M. T.; Bertoni, M. D.; Sepulveda, C. S.; Damonte, E. B.; Maier, M. S. Lichen Secondary Metabolites From the Cultured Lichen Mycobionts of *Teloschistes Chrysophthalmus* and *Ramalina Celastri* and Their Antiviral Activities. *Zeitschr. Naturforsch.* **2007,** *62c*, 543–549.

29. Fernandez-Moriano, C.; Gomez-Serranillos, M. P.; Crespo, A. Antioxidant Potential of Lichen Species and Their Secondary Metabolites. A Systematic Review. *Pharmaceut. Biol.* **2016,** *54* (1), 1–17.

30. Firdous, S. S.; Naz, S.; Shaheen, H.; Dar, M. E. I. Lichens as Bioindicators of Air Pollution From Vehicular Emissions in District Poonch, Azad Jammu and Kashmir, Pakistan. *Pak. J. Bot.* **2017,** *49* (5), 1801–1810.

31. Ganesan, A.; Purushothaman, D. K.; Muralitharan, U.; Subbaiyan, R. Metabolite Profiling and In Vitro Assessment of Antimicrobial and Antioxidant Activities of Lichen *Ramalina Inflata*. *Internat. Res. J. Pharm.* **2016,** *7* (12), 132–138.

32. Geyikoglu, F.; Turk, H.; Aslan, A. The Protective Roles of Some Lichen Species on Colloidal Bismuth Subcitrate Genotoxicity. *Toxicol. Ind. Heal.* **2007,** *23* (8), 487–492.

33. Ghorbani, A.; Langenberger, G.; Sauerborn, J. A Comparison of the Wild Food Plant Use Knowledge of Ethnic Minorities in Naban River Watershed National Nature Reserve, Yunnan, SW China. *J. Ethnobiol. Ethnomed.* **2012,** *8*, 17.

34. Goel, M.; Singh, B. Efficacy of Major Chemical Constituents Isolated From Himalayan Lichen *Ramalina Roesleri* Nyl. against Soil Borne Plant Pathogenic Fungi. *Internat. J. Appl. Engineer. Res.* **2015,** *10* (35), 27490–27495.

35. González, A. G.; Barrera, J. B.; Pérez, E. M.; Padrón, C. E. Chemical Constituents of the Lichen *Ramalina Hierrensis*. *Plant. Med.* **1992,** *58* (2), 214–218.

36. Gonzalez-Tejero, M. R.; Martinez-Lirola, M. J.; Casares-Porcel, M.; Molero-Mesa, J. Three Lichens Used in Popular Medicine in Eastern Andalucia (Spain). *Econ. Bot.* **1995,** *49* (1), 96–98.

37. Goyal, P. K.; Verma, S. K. Pharmacological and Phytochemical Aspects of Lichen *Parmelia Peltata*: A Review. *Internat. J. Res. Ayur. Pharm.* **2016,** *7* (S1), 102–107.

38. Gulluce, M.; Aslan, A.; Sokmen, M.; Sahin, F. Screening the Antioxidant and Antimicrobial Properties of the Lichens *Parmelia Saxatilis*, *Platismatia Glauca*, *Ramalina Pollinaria*, *Ramalina Polymorpha* and *Umbilicaria Nylanderiana*. *Phytomed.* **2006,** *13* (7), 515–521.

39. Gunasekaran, S.; Rajan, V. P.; Ramanathan, S. Antibacterial and Antioxidant Activity of Lichens *Usnea Rubrotincta, Ramalina Dumeticola, Cladonia Verticillata* and Their Chemical Constituents. *Malay. J. Analytic. Sci.* **2016,** *20,* 1–13.

40. Halici, M.; Kufrevioglu, O. I.; Odabasoglu, F. The Ethanol–Water Extract of *Ramalina Capitata* Has Gastroprotective and Antioxidative Properties: An Experimental Study in Rats with Indomethacin-Induced Gastric Injuries. *J. Food Biochem.* **2011,** *35* (1), 11–26.

41. Hanus, L. O.; Temina, M.; Dembitsky, V. Biodiversity of the Chemical Constituents in the Epiphytic Lichenized Ascomycete *Ramalina Lacera* Grown on Difference Substrates *Crataegus Sinaicus, Pinus Halepensis,* and *Quercus Calliprinos. Biomed. Pap. Med. Fac. Univ. Palacky Olomouc Czech. Repub.* 2008, *152* (2), 203–208.

42. Hengameh, P.; Rajkumar, G. H. Assessment of Bactericidal Activity of Some Lichen Extracts by Disc Diffusion Assay. *Internat. J. Drug Develop. Res.* **2017,** *9* (1), 9–19.

43. Hengameh, P.; Rashmi, S.; Garampalli, R. H. *In Vitro* Inhibitory Activity of Some Lichen Extracts Against α-Amylase Enzyme. *Eur. J. Biomed. Pharmaceut. Sci.* **2016,** *3* (5), 315–318.

44. Hoskeri, J. H.; Krishna, V.; Amruthavalli, C. Effects of Extracts From Lichen *Ramalina Pacifica* Against Clinically Infectious Bacteria. *Research.* **2010,** *2* (3), 81–85.

45. Huneck, S. The Significance of Lichens and Their Metabolites. *Naturwissenschaf.* **1999,** *86* (12), 559–570.

46. Kambar, Y.; Vivek, M. N.; Manasa, M.; Kekuda, P. T. R.; Onkarappa, R. Antimicrobial Activity of *Ramalina Conduplicans* Vain. (Ramalinaceae). *Sci. Technol. Arts Res. J.* **2014,** *3* (3), 57–62.

47. Kambar, Y.; Vivek, M. N.; Manasa, M. Antimicrobial Activity of *Leptogium Burnetiae, Ramalina Hossei, Roccella Montagnei* and *Heterodermia Diademata. Internat. J. Pharmaceut. Phytopharmacol. Res.* **2014,** *4* (3), 164–168.

48. Karagöz, A.; Dogruöz, N.; Zeybek, Z.; Aslan, A. Antibacterial Activity of Some Lichen Extracts. *J. Med. Plant. Res.* **2009,** *3* (12), 1034–1039.

49. Karunaratne, V.; Bombuwela, K.; Kathirgamanathar, S.; Thadhani, V. M. Lichens: A Chemically Important Biota. *J. Nat. Sci. Found. Sri Lanka.* **2005,** *33* (3), 169–186.

50. Kekuda, P. T. R.; Vinayaka, K. S. *In Vitro* Anticaries Activity of Some Macrolichens of Karnataka, India. *Internat. J. Pharma Res. Heal. Sci.* **2016,** *4* (3), 1244–1248.

51. Koparal, A. T.; Tüylü, B. A.; Türk, H. *In Vitro* Cytotoxic Activities of (+)-Usnic Acid and (-)-Usnic Acid on V79, A549, and Human Lymphocyte Cells and Their Non-Genotoxicity on Human Lymphocytes. *Nat. Prod. Res.* **2006,** *20* (14), 1300–1307.

52. Kumar, P. S. V.; Kekuda, P. T. R.; Vinayaka, K. S.; Swathi, D.; Chinmaya, A. Insecticidal Efficacy of *Ramalina Hossei* H. Magn and G. Awasthi and *Ramalina conduplicans* Vain macrolichens from Bhadra Wildlife Sanctuary, Karnataka. *Biomed.* **2010,** *30,* 100–102.

53. Kumar, S. V. P.; Kekuda, P. T. R.; Vinayaka, K. S.; Sudharshan, S. J. Anthelmintic and Antioxidant Efficacy of Two Macrolichens of Ramalinaceae. *Pharmacog. J.* **2009,** *1* (4), 238–242.

54. Kumar, S. V. P.; Kekuda, P. T. R.; Vinayaka, K. S.; Swathi, D.; Mallikarjun, N.; Nishanth, B. C. Studies on Proximate Composition, Antifungal and Anthelmintic Activity of a Macrolichen *Ramalina Hossei* H. Magn & G. Awasthi. *Internat. J. Biotechnol. Biochem.* **2010,** *6* (2), 191–201.

55. Lai, D.; Odimegwu, D. C.; Esimone, C.; Grunwald, T.; Proksch, P. Phenolic Compounds With In Vitro Activity Against Respiratory Syncytial Virus From the Nigerian Lichen *Ramalina Farinacea. Planta Medic.* **2013,** *79* (15), 1440–1446.

56. Lal, B.; Upreti, D. K.; Karakoti, B. S. Ethnobotanical Utilization of Lichens by the Tribals of Madhya Pradesh. *J. Econ. Taxonom. Bot.* **1985,** *7,* 203–204.

57. Leão, A. M.; Buchi, D. F.; Iacomini, M.; Gorin, P. A.; Oliveira, M. B. Cytotoxic Effect Against HeLa Cells of Polysaccharides From the Lichen *Ramalina Celastri. J. Submicroscop. Cytol. Pathol.* **1997,** *29* (4), 503–509.

58. Lee, S. G.; Koh, H. Y.; Oh, H.; Han, S. J.; Kim, I. C.; Lee, H. K.; Yim, J. H. Human Dermal Fibroblast Proliferation Activity of Usimine-C From Antarctic Lichen *Ramalina Terebrata. Biotechnol. Lett.* **2010,** *32* (4), 471–475.

59. Lin, C. A. Preliminary Study of the Lichen Genus *Ramalina* at Mt. Yangtou, Hualien County, Eastern Taiwan. *Collect. Res.* **2009,** *22,* 131–134.

60. Liu, B.; Liu, Y.; Li, J.; Gu, R.; Wujisiguleng; Li, P.; Li, F. Aromatic Lichen Resources in Guizhou Province, China. *Med. Arom. Plant.* **2014,** *3,* 146.

61. Luo, H.; Wei, X.; Yamamoto, Y.; Liu, Y.; Wang, L.; Jung, J. S.; Koh, Y. J.; Hur, J. Antioxidant Activities of Edible Lichen *Ramalina Conduplicans* and Its Free Radical-Scavenging Constituents. *Mycosci.* **2010,** *51* (5), 391–395.

62. Molnár, K.; Farkas, E. Current Results on Biological Activities of Lichen Secondary Metabolites—Review. *Zeitschr. Naturforsch.* **2010,** *65c,* 157–173.

63. Moreira, A. S. N.; Braz-Filho, R.; Mussi-Dias, V.; Vieira, I. J. C. Chemistry and Biological Activity of *Ramalina* Lichenized Fungi. *Mol.* **2015,** *20* (5), 8952–8987.

64. Moreira, A. S. N.; Fernandes, R. O. S.; Lemos, F. J. A.; Braz-Filho, R.; Vieira, I. J. C. Larvicidal Activity of *Ramalina Usnea* Lichen against *Aedes Aegypti. Brazil. J. Pharmacog.* **2016,** *26,* 530–532.

65. Nanayakkara, C.; Bombuwela, K.; Kathirgamanathar, S. Effect of Some Lichen Extracts from Sri Lanka on Larvae of *Aedes Aegypti* and the Fungus *Cladosporium Cladosporioides. J. Nat. Sci. Found. Sri Lanka.* **2005,** *33* (2), 147–149.

66. Odimegwu, D. C.; Okore, V. C.; Esimone, C. O. Analyses of *Ramalina Farinaceae* Extract Anti-Proliferative Activities in Culture of Mammalian Cells and Splenocytes. *Res. J. Med. Plant.* **2016,** *10* (4), 303–308.

67. Oh, S.; Wang, X. Y.; Wang, L. S.; Liu, P. G.; Hur, J. A Note on the Lichen Genus *Ramalina* (Ramalinaceae, Ascomycota) in the Hengduan Mountains in China. *Mycobiol.* **2014,** *42* (3), 229–240.

68. Oksanen, I. Ecological and Biotechnological Aspects of Lichens. *Appl. Microbiol. Biotechnol.* **2006,** *73,* 723–734.

69. Olafsdottir, E. S.; Ingólfsdottir, K. Polysaccharides From Lichens: Structural Characteristics and Biological Activity. *Plant. Medic.* **2001,** *67* (3), 199–208.

70. Özyiğitoğlu, G. C.; Açikgöz, B.; Sesal, C. Lichen Secondary Metabolites: Synthesis Pathways and Biological Activities. *Acta Biologic. Turc.* **2016,** *29* (4), 150–163.

71. Parizadeh, H.; Garampalli, R. H. Evaluation of Some Lichen Extracts for β-Glucosidase Inhibitory as a Possible Source of Herbal Anti-Diabetic Drugs. *Am. J. Biochem.* **2016,** *6* (2), 46–50.

72. Parizadeh, H.; Garampalli, R. H. Physiological and Chemical Analysis for Identification of Some Lichen Extracts. *J. Pharmacog. Phytochem.* **2017,** *6* (5), 2611–2621.

73. Park, B.; Yim, J. H.; Lee, H. K.; Kim, B. O.; Pyo, S. Ramalin Inhibits VCAM-1 Expression and Adhesion of Monocyte to Vascular Smooth Muscle Cells Through MAPK and PADI4-Dependent NF-kB and AP-1 Pathways. *Biosci. Biotechnol. Biochem.* **2015,** *79* (4), 539–552.

74. Park, H. J.; Jang, Y. J.; Yim, J. H.; Lee, H. K.; Pyo, S. Ramalin Isolated From *Ramalina Terebrata* Attenuates Atopic Dermatitis-Like Skin Lesions in Balb/c Mice and Cutaneous Immune Responses in Keratinocytes and Mast Cells. *Phytother. Res.* **2016,** *30* (12), 1978–1987.

75. Park. S.; Jang, S.; Oh, S.; Kim, J. A.; Hur, J. An Easy, Rapid, and Cost-Effective Method for DNA Extraction From Various Lichen Taxa and Specimens Suitable for Analysis of Fungal and Algal Strains. *Mycobiol.* **2014,** *42* (4), 311–316.

76. Parrot, D.; Jan, S.; Baert, N.; Guyot, S.; Tomasi, S. Comparative Metabolite Profiling and Chemical Study of *Ramalina Siliquosa* Complex Using LC–ESI-MS/MS Approach. *Phytochem.* **2013,** *89,* 114–124.

77. Pathak, A.; Mishra, R. K.; Shukla, S. K.; Kumar, R.; Pandey, M.; Pandey, M.; Qidwai, A.; Dikshit, A. In Vitro Evaluation of Antidermatophytic Activity of Five Lichens. *Cog. Biol.* **2016,** *2,* 1197472.

78. Paudel, B.; Bhattarai, H. D.; Koh, H. Y.; Lee, S. G. *Ramalina*—A Novel Nontoxic Antioxidant Compound From the Antarctic Lichen *Ramalina Terebrata*. *Phytomed.* **2011,** *18* (14), 1285–1290.

79. Paudel, B.; Bhattarai, H. D.; Lee, H. K. Antibacterial Activities of Ramalin, Usnic Acid and Its Three Derivatives Isolated From the Antarctic Lichen *Ramalina Terebrata*. *Zeitsch. Naturforsch.* **2010,** *65c,* 34–38.

80. Paudel, B.; Bhattarai, H. D.; Lee, J. S.; Hong, S. G.; Shin, H. W.; Yim, J. H. Antibacterial Potential of Antarctic Lichens Against Human Pathogenic Gram-Positive Bacteria. *Phytother. Res.* **2008,** *22* (9), 1269–1271.

81. Quedensley, T. S.; Perez, M. V. *Ramalina Mahoneyi*—A New Corticolous Lichen From a Western Guatemalan Cloud Forest. *Lundellia* **2011,** *14,* 3–7.

82. Ramya, K.; Thirunalasundari, T. Evaluation of Antioxidant Activity Among the Epiphytic Lichens of Kodaikanal. *World J. Pharm. Pharm. Sci.* **2017,** *6* (11), 828–841.

83. Redzic, S.; Barudanovic, S.; Pilipovic, S. Wild Mushrooms and Lichens Used As Human Food for Survival in War Conditions; Podrinje–Zepa region (Bosnia and Herzegovina, W. Balkan). *Res. Hum. Ecol.* **2010,** *17* (2), 175–187.

84. Ristic, S.; Rankovic, B.; Kosanić, M. Biopharmaceutical Potential of Two *Ramalina* Lichens and Their Metabolites. *Curr. Pharm. Biotechnol.* **2016,** *17* (7), 651–658.

85. Saha, D.; Sundriyal, R. C. Perspectives of Tribal Communities on NTFP Resource Use in a Global Hotspot: Implications for Adaptive Management. *J. Nat. Sci. Res.* **2013,** *3* (4), 125–169.

86. Saklani, A.; Upreti, D. K. Folk Uses of Some Lichens in Sikkim. *J. Ethnopharm.* **1992,** *37,* 229–233.

87. Santos, L. S.; Soriano, M. P. C.; Mirabal-Gallardo, Y. Chemotaxonomic Fingerprinting of Chilean Lichens Through Maldi and Electrospray Ionization Mass Spectrometry. *Brazil. Arch. Biol. Technol.* **2015,** *58* (2), 244–253.

88. Sesal, C.; Çobanoğlu, G.; Karaltı, İ.; Açıkgöz, B. *In Vitro* Antimicrobial Potentials of Four *Ramalina* Lichen Species from Turkey. *Curr. Res. Environ. Appl. Mycol.* **2016,** *6* (3), 202–209.

89. Shah, N. C. Lichens of Commercial Importance in India. *Scit. J.* **2014,** *1* (2), 32–36.

90. Shivanna, R.; Garampalli, R. H. Investigation of Macrolichens for Antifungal Potentiality Against Phytopathogens. *Ind. Am. J. Pharm. Res.* **2016,** *6* (4), 5290–5296.

91. Shivanna, R.; Parizadeh, H.; Garampalli, R. H. *In Vitro* Anti-Obesity Effect of Macrolichens *Heterodermia Leucomelos* and *Ramalina Celastri* by Pancreatic Lipase Inhibitory Assay. *Int. J. Pharm. Pharm. Sci.* **2017,** *9* (5), 137–140.

92. Shrestha, G.; Raphael, J.; Leavitt, S. D.; St. Clair, L. L. *In Vitro* Evaluation of the Antibacterial Activity of Extracts From 34 Species of North American Lichens. *Pharm. Biol.* **2014,** *52* (10), 1262–1266.

93. Shrestha, G.; St. Clair, L. L. Antimicrobial Activity of Extracts from Two Lichens *Ramalina Menziesii* and *Usnea Lapponica. Bull. Calif. Lichen Soc.* **2013,** *20* (1), 5–10.

94. Shrestha, G.; Thompson, A.; Robison, R.; St. Clair, L. L. *Letharia Vulpina,* A Vulpinic Acid Containing Lichen, Targets Cell Membrane and Cell Division Processes in Methicillin-Resistant *Staphylococcus Aureus. Pharm. Biol.* **2015,** *54* (3), 413–418.

95. Shukla, V.; Joshi, G. P.; Rawat, M. S. M. Lichens as a Potential Natural Source of Bioactive Compounds—Review. *Phytochem. Rev.* **2010,** *9* (2), 303–314.

96. Shukla, V.; Negi, S.; Rawat, M. S. M.; Pant, G.; Nagatsu, A. Chemical Study of *Ramalina Africana* (Ramalinaceae) from the Garhwal Himalayas. *Biochem. Syst. Ecol.* **2004,** *32,* 449–453.

97. Sigdel, S. R.; Rokaya, M. B.; Timsina, B. Plant Inventory and Ethnobotanical Study of Khimti Hydropower Project, Central Nepal. *Sci. World* **2013,** *11* (9), 105–112.

98. Sisodia, R.; Goel, M.; Verma, S.; Rani, A.; Dureja, P. Antibacterial and Antioxidant Activity of Lichen Species *Ramalina Roesleri. Nat. Prod. Res.* **2013,** *27* (23), 2235–2239.

99. Smitha, K. C.; Garampalli, R. H. Evaluation of Phytochemicals and *In Vitro* Antioxidant Activity of *Ramalina Pacifica* and *Roccella Montagnei. J. Pharm. Phytochem.* **2016,** *5* (2), 270–274.

100. Stanly, C.; Ali, D. M. H.; Keng, C. L.; Boey, P.; Bhatt, A. Comparative Evaluation of Antioxidant Activity and Total Phenolic Content of Selected Lichen Species from Malaysia. *J. Pharm. Res.* **2011,** *4* (8), 2824–2827.

101. Stark, J. B.; Walter, E. D.; Owens, H. S. Method of Isolation of Usnic Acid From *Ramalina Reticulata. J. Am. Chem. Soc.* **1950,** *72* (4), 1819–1820.

102. Suh, S.; Kim, T. K.; Kim, J. E.; Hong, J.; Nguyen, T. T. T.; Han, S. J.; Youn, U. J.; Yim, J. H.; Kim, I. Anticancer Activity of Ramalin, A Secondary Metabolite From the Antarctic Lichen *Ramalina Terebrata,* against Colorectal Cancer Cells. *Mol.* **2017,** *22,* 1361–1367.

103. Suyanto, I. Silico Analysis of Depsidone as Anticancer. *Der. Pharm. Chem.* **2017,** *9* (11), 5–8.

104. Svanberg, I.; Egisson, S. Edible Wild Plant Use in the Faroe Islands and Iceland. *Acta. Soc. Botanic. Pol.* **2012,** *81* (4), 233–238.

105. Tabbabi, K.; Karmous, T. Characterization and Identification of the Components Extracted From 28 Lichens in Tunisia by High Performance Thin-Layer Chromatography (HPTLC), Morphologic Determination of the Species and Study of the Antibiotic Effects of Usnic Acid. *Med. Arom. Plant.* **2016,** *5,* 253.

106. Tang, Y.; Laxi, N.; Bao, H. Chemical Constituents and Their Antioxidant Activities From Thallus of *Ramalina Conduplicans. Mycosyst.* **2015,** *34* (1), 169–176.

107. Tay, T.; Türk, A. O.; Yılmaz, M. Evaluation of the Antimicrobial Activity of the Acetone Extract of the Lichen *Ramalina Farinacea* and Its ±Usnic Acid, Norstictic Acid, and Protocetraric Acid Constituents. *Zeitsch. Naturforsch.* **2004,** *59c,* 384–388.

108. Timbreza, L. P.; Delos Reyes, J. L.; Flores, C. H. C. Antibacterial Activities of the Lichen *Ramalina* and *Usnea* Collected from Mt. Banoi, Batangas and Dahilayan, Bukidnon, against Multi-Drug Resistant (MDR) Bacteria. *Aust. J. Mycol.* **2017,** *26,* 27–42.

109. Upreti, P. K.; Divakar, D. K.; Nayaka, S. Commercial and Ethnic Uses of Lichens in India. *Econ. Bot.* **2005,** *59* (3), 269–273.

110. Valadbeigi, T.; Shaddel, M. Amylase Inhibitory Activity of Some Macrolichens in Mazandaran Province, Iran. *Physiol. Pharmacol.* **2016**, *20*, 215–219.

111. Vattem, D. A.; Vaden, M.; Jamison, B. Y.; Maitin, V. Antioxidant and Anti-Adhesive Activity of Some Common Lichens. *J. Pharmacol. Toxicol.* **2012**, *7*, 96–103.

112. Verma, N.; Behera, B. C.; Sharma, B. O. Glucosidase Inhibitory and Radical Scavenging Properties of Lichen Metabolites Salazinic Acid, Sekikaic Acid and Usnic Acid. Hacettepe *J. Biol. Chem.* **2012**, *40* (1), 7–21.

113. Vinayaka, K. S.; Karthik, S.; Nandini, K. C.; Kekuda, P. T. R. Amylase Inhibitory Activity of Some Macrolichens of Western Ghats, Karnataka, India. *Ind. J. Novel Drug Deliv.* **2013**, *5* (4), 225–228.

114. Vinayaka, K. S.; Kekuda, P. T. R.; Swathi, D.; Kumar, P. S. V. Studies on Chemical Composition and *In Vitro* Antibacterial Activity of Solvent Extracts of Lichen *Ramalina Hossei* Vain. (Ramalinaceae). *Biotechnol. Ind. J.* **2009**, *3* (4), 309–311.

115. Vinayaka, K. S.; Krishnamurthy, Y. L. Ethno-Lichenological Studies of Shimoga and Mysore Districts, Karnataka, India. *Adv. Plant Sci.* **2012**, *25*, 265–267.

116. Vinayaka, K. S.; Shetty, S. R.; Krishnamurthy, Y. L. Utilization of Lichens in the Central Western Ghats Area of Karnataka, India. *Brit. Lichenolog. Soc. Bull.* **2011**, *109*, 56–62.

117. Wang, L.; Narui, T.; Harada, H.; Culberson, C. F.; Culberson, W. L. Ethnic Uses of Lichens in Yunnan, China. *Bryol.* **2001**, *104* (3), 345–349.

118. Weissbuch, B. K. Medicinal Lichens: The Final Frontier. *J. Am. Herbal. Guild* **2016**, *12* (2), 23–28.

119. Zrnzević, I.; Jovanović, O.; Zlatanović, I.; Stojanović, I. Constituents of *Ramalina Capitata* (Ach.) Nyl. Extracts. *Nat. Prod. Res.* **2017**, *31* (7), 857–860.

120. Zrnzevic, I.; Stankovic, M.; Jovanovic, V. S.; Mitic, V. *Ramalina Capitata* (Ach.) Nyl. Acetone Extract: HPLC Analysis, Genotoxicity, Cholinesterase, Antioxidant and Antibacterial Activity. *Excli. J.* **2017**, *16*, 679–687.

APPENDIX I: CLASSIFICATION OF LICHENS.

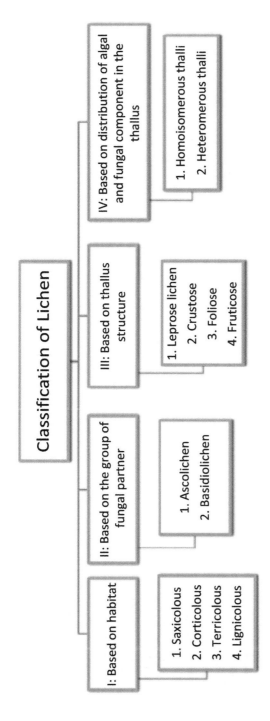

Classification of Lichen

I: Based on habitat

1. Saxicolous
2. Corticolous
3. Terricolous
4. Lignicolous

II: Based on the group of fungal partner

1. Ascolichen
2. Basidiolichen

III: Based on thallus structure

1. Leprose lichen
2. Crustose
3. Foliose
4. Fruticose

IV: Based on distribution of algal and fungal component in the thallus

1. Homoisomerous thalli
2. Heteromerous thalli

Source: www.plantscience4u.com.

APPENDIX II: LIST OF RAMALINA SPECIES

Ramalina ahtii
Ramalina alludens
Ramalina americana
Ramalina anceps
Ramalina asahinae
Ramalina aspera
Ramalina asperula
Ramalina bajacaliformica
Ramalina baltica
Ramalina caespitella
Ramalina calcarata
Ramalina camptospora
Ramalina candicularis
Ramalina canariensis
Ramalina capensis
Ramalina capitata
Ramalina celastri
Ramalina chihuahuana
Ramalina chondrina
Ramalina cochlearis
Ramalina complanata
Ramalina cryptochlorophaea
Ramalina culbersoniorum
Ramalina cuspidata
Ramalina darwiniana
Ramalina dendriscoides
Ramalina dendroides
Ramalina dilacerata

Ramalina erosa
Ramalina exiguella
Ramalina farinacea
Ramalina fastigiata
Ramalina flavovirens
Ramalina fragilis
Ramalina fraxinea
Ramalina furcellangulida
Ramalina grumosa
Ramalina hyrcana
Ramalina intermedia
Ramalina jamesii
Ramalina lacera
Ramalina leptocarpha
Ramalina litorea
Ramalina mahoneyi
Ramalina menziesii
Ramalina meridionalis
Ramalina microspora
Ramalina montana
Ramalina moranii
Ramalina palmiformis
Ramalina peruviana
Ramalina pollinaria
Ramalina polyforma
Ramalina polymorpha
Ramalina portosantana
Ramalina portuensis

Ramalina prolifera
Ramalina psoropmica
Ramalina puberulenta
Ramalina puiggarii
Ramalina pusiola
Ramalina quercicola
Ramalina rectangularis
Ramalina rapestris
Ramalina sharpii
Ramalina siliquosa
Ramalina sinaloensis
Ramalina sinensii
Ramalina sonorensis
Ramalina sorediosa
Ramalina sprengelii
Ramalina stenospora
Ramalina stoffersii
Ramalina subcalicaris
Ramalina subfraxinea
Ramalina subleptocarph
Ramalina superfraxinea
Ramalina thrausta
Ramalina timdaliana
Ramalina usnea
Ramalina whinrayi
Ramalina wigginsiae
Ramalina willeyi

Source: https://en.wikipedia.org/wiki/Ramalina.

APPENDIX III: LIST OF COMMON NAMES OF LICHENS.

Ball lichen – *Sphaerophorus*
Barnacle lichen – *Thelotrema*
Beard lichen – *Usnea*
Birch bark dot lichen – *Leptorhaphis*
Black curly lichen – *Pseudephebe*
Black thread lichen – *Placynthium*
Blemished lichen – *Phlyctis*
Blistered navel lichen – *Lasallia*
Blood lichen – *Mycoblastus*
Blood stain lichen – *Haematomma*
Bowl lichen – *Psoroma*
Bran lichen – *Parmeliopsis*
Brittle lichen – *Cornicularia*
Bruised lichen – *Toninia*
Bulls eye lichen – *Placopsis*
Button lichen – *Buellia*

Cap lichen – *Baeomyces*
Cartilage lichen – *Ramalina*
Chalice lichen – *Endocarpon*
Chocolate chip lichen – *Solorina*
Clam lichen – *Normandina*
Club lichen – *Multiclavula*
Cobble stone lichen – *Acarospora*
Cockle shell lichen – *Hypocenomyce*
Comma lichen – *Arthonia*
Coral lichen – *Sphaerophorus*
Crabs eye lichen – *Ochrolechia*
Cracked lichen – *Acarospora*
Crater lichen – *Diploschistes*
Cup lichen – *Cladonia*

Dark lichen – *Pseudevernia*
Dimple lichen – *Gyalecta*
Disc lichen – *Buellia*
Disk lichen – *Lecidea*
Disk lichen – *Trapelia*
Disk lichen – *Tremolecia*
Dot lichen – *Arthonia*
Dot lichen – *Micarea*
Dotted lichen – *Bacidia*
Dust lichen – *Chrysothrix*
Dust lichen – *Lepraria*

Earth lichen – *Catapyrenium*
Egg yolk lichen – *Candelariella*
Felt lichen – *Peltigera*
Fire dot lichen – *Caloplaca*
Fish scale lichen – *Psora*
Fringed lichen – *Anaptychia*
Frosted lichen – *Physconia*

Giant shield lichen – *Cetrelia*
Gold dust lichen – *Chrysothrix*
Gold lichen – *Caloplaca*
Gold speck lichen – *Candelariella*
Granular lichen – *Lopadium*

Honey combed lichen – *Menegazzia*
Horse hair lichen – *Bryoria*
Jelly lichen – *Collema*
Jewel lichen – *Caloplaca*
Kidney lichen – *Nephroma*

Lead lichen – *Parmeliella*
Light lichen – *Pseudevernia*
Lung lichen – *Lobaria*

Map lichen – *Rhizocarpon*
Matted lichen – *Pannaria*
Mearly lichen – *Leprocaulon*
Moon glow lichen – *Dimelaena*

Nail lichen – *Pilophorus*
Navel lichen – *Umbilicaria*
Needle lichen – *Chaenotheca*

Old wood rimmed lichen – *Lecanactis*
Orange lichen – *Caloplaca*
Orange lichen – *Xanthomendoza*
Orange lichen – *Xanthoria*
Orange wall lichen – *Xanthoria*

Peppermint drop lichen – *Icmadophila*
Pin lichen – *Calicium*
Pin lichen – *Chaenotheca*
Pore lichen – *Pertusaria*
Puffed sunken disk lichen – *Lobothallia*

Ragged lichen – *Platismatia*
Reindeer lichen – *Cladina*
Rim lichen – *Lecanora*
Rim lichen – *Squamarina*
Rimmed lichen – *Aspicilia*
Rimmed navel lichen – *Rhizoplaca*
Ring lichen – *Evernia*
Rock olive lichen – *Peltula*
Rock pimple lichen – *Staurothele*
Rock posy lichen – *Rhizoplaca*
Rock shield lichen – *Xanthoparmelia*
Rock tripe lichen – *Umbilicaria*
Rosette lichen – *Physcia*
Ruffle lichen – *Parmotrema*

Scatter rug lichen – *Parmotrema*
Scribble lichen – *Opegrapha*
Script lichen – *Graphis*
Sea weed lichen – *Lichina*
Shell lichen – *Arthopyrenia*
Shield lichen – *Heterodermia*
Shield lichen – *Parmelia*
Silver skin lichen – *Dermatocarpon*
Skin lichen – *Leptogium*
Snow lichen – *Stereocaulon*
Socket lichen – *Solorina*
Soot lichen – *Cyphelium*

Spike lichen – *Calicium*
Spotted felt lichen – *Sticta*
Stipple back lichen – *Dermatocarpon*
Stipple scale lichen – *Placidium*
Strap lichen – *Ramalina*
Strap lichen – *Calicium*
Sulphur lichen – *Fulgensia*
Sun burst lichen – *Xanthomendoza*
Sun burst lichen – *Xanthoria*
Sunken disk lichen – *Aspicilia*

Thread lichen – *Ephebe*
Tile lichen – *Lecidea*
Tube lichen – *Hypogymnia*
Urn lichen – *Tholurna*
Velvet lichen – *Cystocoleus*

Wart lichen – *Pertusaria*
Wart lichen – *Porina*
Wart lichen – *Pyrenula*
Wart lichen – *Staurothele*
Wart lichen – *Verrucaria*
White fingers lichen – *Siphula*
White worm lichen – *Thamnolia*
Witch's hair lichen – *Alectoria*
Wolf lichen – *Letharia*
Wreath lichen – *Phaeophyscia*

Source: http://eol.org/pages/133441/overview; http://plants.usda.gov/java/name Search; https://www.britannica.com/topic/list-of-lichens-2032425; Sharnoff, S. *Field Guide to California Lichens*. Yale University Press, 2014; ISBN 978-0-300-19500-2.

CHAPTER 14

ROLE OF ANTIOXIDANT-BASED NUTRACEUTICALS IN TRANSLATIONAL MEDICINE: REVIEW

HARISHKUMAR MADHYASTHA, RADHA MADHYASTHA, YUTTHANA PENGJAM, YUICHI NAKAJIMA, and MASUGI MARUYAMA

ABSTRACT

Medicinal plants possess curative properties due to their ability to inhibit oxidants and prevent oxidative stress, besides stimulating the activation of factors to promote human health. Research studies have highlighted the potential of polyphenols in modulating various molecular processes involved in oxidant activation in several disease conditions. A combination of these natural agents with traditional chemo-based or hormone-based therapies opens up a novel approach to fight against diseases. This chapter discusses molecular systems that link polyphenols to inhibition of oxidative stress occurring in pathological events such as cancer, psychiatric and cognitive disorder, cardiovascular disease, and neurological memory loss disease. The objective of this review is to increase awareness of the importance of dietary choices as palliative measures of disease prevention.

14.1 INTRODUCTION

Medicinal plants have a long history of acknowledgment as agents with beneficial effects on biological functions, mainly through their antioxidant and anti-inflammatory activities. There exists an inverse relationship between antioxidant-rich food consumption and disease development. They have a potential to improve the lives of patients suffering from cancer, cardiovascular, and Alzheimer's disease. Their physiological efficiency is attributed to the presence of phytochemicals, especially polyphenols that include flavonoids (quercetin, catechin, anthocyanin), lignans, phenolic acids (vanillic acid, caffeic acid), stilbenes (resveratrol), and others, such as, curcumin.[41] In recent few decades, rise in awareness regarding health benefits of polyphenols has been observed. There is substantial increase in the amount of research carried out worldwide, to reveal the molecular mechanisms responsible for the medicinal properties of polyphenols.[8,27,28,29,41]

In this chapter, authors have highlighted the impact of plant-based polyphenols on pathological human conditions, such as psychiatric and cognitive disorders, cardiovascular disease, cancer, and Alzheimer's disease.

14.2   ROLE OF OXIDATIVE STRESS IN PHYSIOLOGICAL PROCESSES

Metabolism of oxygen in the cells due to cellular respiration can generate potentially deleterious oxidants, called reactive oxygen species (ROS). ROS include free radicals (such as OH^- (hydroxyl), O_2^- (superoxide), and $ONOO^-$ (peroxynitrite)), and non-radicals species (such as H_2O_2 (hydrogen peroxide)). ROS are generated as byproducts of metabolic processes through activation of enzymes, such as peroxidase, nitric oxide synthase (NOS), NADPH oxidase, cyclooxygenase, lipoxygenase, myeloperoxidase, xanthine oxidase, and cytochrome P_{450} monooxygenase.[5] Physiologically, ROS generation in moderate level, is vital to maintain physiological activity and normal cellular processes including host defense, and healthy cellular homeostasis, through their ability to modulate redox signaling molecules.[5] ROS regulate cellular processes, such as proliferation, migration, differentiation, cell cycle, cytoskeleton reorganization, through activation of several signaling pathways, including mitogen-activated protein kinase (MAPK), tyrosine kinase, Rho kinase, and transcription factors such as NFκB, AP-1, and HIF-1.[51] ROS can also upregulate expression and activity of oncogenes and pro-inflammatory genes.[26]

Under normal conditions, the rate and magnitude of formation of oxidants is subdued by the rate of elimination of oxidants by antioxidants. An imbalance in the ratio of oxidants to antioxidants is due to uncontrolled generation of ROS target to oxidative stress and consequent damage of DNA, proteins and lipids, resulting in cell injury and cell death.[37] Human body has powerful antioxidant defense systems comprising of enzymatic and nonenzymatic components to maintain the oxidant–antioxidant balance. Enzymatic antioxidants are manganese superoxide dismutase (MnSOD), catalase, glutathione peroxidase, thioredoxin, and peroxiredoxin. Nonenzymatic antioxidants include albumin, bilirubin, uric acid, and glutathione. Micronutrients, and vitamins A, C, and E also form the defense line against oxidative stress.[37] However, human body cannot manufacture these micronutrients, and hence, dietary intervention becomes necessary.

14.3   ANTIOXIDANTS: DEFINITIONS AND FUNCTIONS

Halliwell and Gutteridge[8] defined antioxidants as *"any substance that when present at low concentrations compared with that of an oxidizable substrate, significantly delays or inhibits oxidation of that substrate."* An effective antioxidant functions at a concentration ratio of 1:100 in relation to the free radicals.[18] Antioxidants can delay or inhibit the chain reaction of oxidation by scavenging the radical by virtue of being an electron donor. They function at three principal levels of defense–prevention, interception, and repair.[37]

Antioxidants can prevent generation of oxidative stress through enzymatic reactions and metal chelation. They can intercept and inhibit chain reactions of oxidation through metal chelation, deactivation, or delocalization of the reactive species. Finally, they can also repair the damage caused by oxidative stress, through DNA repair and proteolytic enzymes that can recognize and remove oxidative modified proteins. Many phytochemicals (especially polyphenols) are recognized as effective antioxidants due to their free radicals or active oxygen scavenging properties.

14.4   POLYPHENOLS

Polyphenols are a class of plant-derived organic compounds that are produced as defense mechanisms against pathogens and physical damage.[27] They are composed of phenolic groups associated with carbohydrates and organic acids. They can be chemically characterized into different synthetic groups based on number of phenol rings they possess and the substances attached to the phenol rings. Flavonoids are most commonly observed polyphenolic groups, which comprise of subgroups, such as anthocyanins, isoflavanoids, flavanols, flavanones, and flavones.

Polyphenols are available as esters, glycosides, and polymers that cannot be absorbed by the gut. They are first hydrolyzed by intestinal enzymes, and conjugated either by methylation, sulfation, or glucuronidation, following which they are absorbed in the gut and carried to different tissues and organs.[49]

Polyphenols function as antioxidants by donating hydrogen atoms from their hydroxyl groups, and generating phenoxyl radical that will donate another hydrogen, or react with another radical to form a stable compound.[32] For example, polyphenol (X-OH) donates a hydrogen atom to the reactive species (R) through hemolytic breaking of the hydroxyl bond. The products, R-H (stabilized oxygen species), and phenoxyl radicals can then be stabilized by reactions with other radicals.[31]

14.5 MOLECULAR MECHANISMS OF POLYPHENOLS AGAINST DISEASES

14.5.1 CANCER

Studies from various groups have successfully documented the beneficial aspects of polyphenols in preventing and treating various kinds of cancers.[2,42,43,49,52] The most noteworthy examples include epigallocatechin gallate (EGCG), curcumin, and genistein, among others. EGCG is predominantly found in green tea, and it inhibits cellular transformation and proliferation, thus protecting from tumorigenesis and malignancy.[38] It also exhibits anticancer effect by arresting the cell cycle in G1 phase,[33,42] and by influencing synthesis of cell cycle progression regulators, CDK1, cyclins D and E. It can also induce apoptosis of cancer cells, by positively regulating the function of proteins, such as: pRd, p53, and p27,[9] and downregulating Bcl-2α and Bcl-xl through nuclear factor-κB (NFκB) inactivation.[25]

NFκB, a known key player in the inflammatory process, is a major signaling pathway that is usually deregulated in cancer pathology, and is actively involved in progression of the disease. NFκB is constitutively active in ER-negative breast cancer cells, due to over activation of Her2/Neu pathway. EGCG was able to reduce expression and activation of NFκB in ER-negative breast cancer cells, by inhibiting Her-2/Neu signaling pathway.[30]

Curcumin, a polyphenol derived from plant turmeric, has a very long history of application in traditional Asian medicine due to its anti-inflammatory, antimicrobial, and antioxidant properties.[6,12,20] Curcumin is poorly absorbed in the gut and has a low rate of bioavailability since it is quickly metabolized by conjugation. However, its high modulatory efficiency has led researchers to enhance its bioavailability and or develop synthetic analogs to promote its biological effects. Primary target of curcumin includes NFκB.[14] The influential effect of curcumin on NFκB also extends to downstream members, such as COX2, MMP2, Bcl-xL, and Bcl2. In addition, studies suggest that curcumin may also have inhibitory effects on other oncogenic signaling pathways, such as Akt, culminating in initiation of apoptotic processes and inhibition of tumor cell proliferation.[45,48]

Genistein is an isoflavone that is abundantly found in soybeans and in most products containing soy protein. Genistein is a negative regulator of signaling molecules that are involved in control of cell cycle. And it also helps in regulation of apoptosis mechanism of cancer cells. It induces growth inhibition and apoptotic cell death in androgen unresponsive prostate cancer cells by inhibiting Akt/ FOXO3a/ GSK-3β/ AR and Akt/ mTOR signaling networks through suppression of tyrosine kinase activity.[2] Genistein has structural similarity to estradiol and can bind to estrogen receptors, α and β. It can inhibit cell growth and proliferation of breast and prostate cancer cells.[21, 34] Additionally, it also displays antioxidant and anticarcinogenic effects through inhibition of NFκB activation.[3,17] Other important targets of genistein include Akt kinase, Wnt/ β-catenin, and Notch signaling.[1,21,44] In addition to these carcinogenic pathways, genistein also directly regulates expression of several other genes that are involved in DNA damage repair, such as GPx, GADD45, glutathione reductase, and GST1, thereby offering protection against oxidative stress and carcinogenesis.[21]

14.5.2 PSYCHIATRIC AND COGNITIVE DISORDER

ROS-induced oxidative stress plays an important role in cognitive decline and neurodegenerative diseases. ROS in brain is induced by the production of nitric oxide radical through increase in expressions of iNOS and nNOS.[7]

Polyphenols exhibit neuroprotective effects through control of four major signaling pathways that are involved with the cognitive processes and neuronal plasticity, most notably PI3K/ Akt, NFκB, CREB, and mitogen-activated protein kinase (MAPK).[10, 39] Animal experiments reveal the efficacy of polyphenols in promoting cognitive function through governing of extracellular signal-regulated kinase (ERK)/CREB signaling pathway and protection from neuro-inflammation.[40] Aged rats on 2% blueberry supplement diet for 12 weeks demonstrated improved cognitive function, which was attributed to activation of CREB protein and high expression of brain-derived neurotrophic factor in the hippocampus.[47]

High consumption rate of green tea among Japanese population is associated with low prevalence of cognitive impairment in the elderly people.[15] The neuroprotective feature of green tea is attributed to NFκB activation suppression by EGCG and L-theanine.[13] These studies indicate powerful neuroprotective properties of polyphenols, which function through multiple routes targeting oxidative stress and cytotoxicity.

14.5.3 CARDIOVASCULAR DISEASES

Diets rich in polyphenols have long been associated with protection from cardiovascular diseases. Several studies, in vitro and in vivo, emphasize the vascular-protective properties of polyphenols. Vascular impairment results from an imbalance between vasorelaxation and vasoconstriction, caused by reduced NO and increased NADPH oxidase-dependent oxidative stress. Quercetin and catechins were effective in reducing vascular superoxide, leukotriene B4, and plasma P-selectin activity and while simultaneously increasing vascular eNOS activity and heme oxygenase-1 protein, in atherosclerotic rats.[19] Mediterranean diet rich in polyphenols helps to protect from cardiovascular disease through their inhibitory effects on intracellular ROS levels and inflammatory pathways involving NFκB.[35] NFκB inhibition causes a decrease in matrix metalloproteinase-9 expression, culminating in inhibition of angiogenesis of endothelial cells. Clinical studies reveal that daily consumption of Mediterranean diet (comprising of olive oil and nuts) reduces the expression of pro-inflammatory cytokines through disruption of TNFα signaling, thus protecting from plaque formation.[4] Widmer and others reported that daily consumption of 30 ml of olive oil for four months improved the endothelial function of patients with early atherosclerotic endothelial dysfunction, which was associated with a reduction in serum ICAM activity.[46]

Epicatechin is also effective in reducing oxidative stress through modulation of VCAM-1 and ICAM-1 in mice model.[24] Resveratrol protects the cardiovascular system through multiple pathways, such as upregulation of thioredoxin, a protein reductase that is necessary to maintain redox environment in the myocytes,[41] besides upregulation of superoxide dismutase, catalase, and glutathione peroxidases via activation of NRF2 signal pathway.

14.5.4 ALZHEIMER'S DISEASE

Alzheimer's disease (AD) is an old age neurodegenerative disorder characterized by deposition of extracellular amyloid beta polypeptide (Aβ) plaques and intracellular neurofibrillary tangles (NFT). Accumulation of Aβ aggregates leads to oxidative stress, inflammation and memory loss, and cognitive dysfunction. EGCG of green tea can rescue from memory impairments induced by Aβ, through inhibition of NFκB pathway.[16] Anthocyanin fraction of blueberry extract also protects against Aβ-induced neurotoxicity through anti-inflammatory and antioxidant properties.[11]

Resveratrol, a polyphenol found in grapes, is shown to protect neurons against Aβ-induced oxidative stress and neurotoxicity through modulation of protein kinase-C.[32] Curcumin also protects cells from Aβ-induced cytotoxicity through antioxidant pathway.[32] Supplementation of food ingredients by curcumin with ferulic acid can reduce oxidative damage and amyloid pathology in AD.[23] Studies suggest that long-term intake of ferulic acid could delay the progression of AD.[36] Yan et al. also reported that long-term administration of ferulic acid protected mice against Aβ-induced learning and memory dysfunction.[50] Regular intake of polyphenol-rich diet is believed to attenuate the progression of AD and also protect the brain from Aβ neurotoxicity.

14.6 SUMMARY

The antioxidant-based nutraceuticals are supportive therapeutic means to promote human health and prevent pathological conditions. Polyphenols exert their positive effects on human health by negatively modulating disease-promoting molecules and enhancing health-promoting factors through their antioxidant properties. However, most of these promising results are derived either from in vitro, animal models, cohort, or small-sized human studies. Research data relating to long-term randomized human studies is lacking. Another major factor stalling their pharmaceutical application is the bioavailability of these natural compounds. Most of these natural products cannot be readily absorbed by the intestine in their native form. Despite these issues, the biological efficacy of nutraceuticals as antioxidants and health promoters has been recorded from ancient days with no known ill effects, and hence, offer great hope as dietary sources of good health.

KEYWORDS

- antioxidant
- cardiovascular disease
- cognitive disorders
- nutraceuticals
- oxidative stress
- phytochemicals
- polyphenols

REFERENCES

1. Anastasius, N.; Boston, S.; Lacey, M.; Storing, N.; Whitehead, S. A. Evidence that Low-Dose, Long-Term Genistein Treatment Inhibits Estradiol-Stimulated Growth in MCF-7 Cells by Down-Regulation of the PI3-Kinase/Akt Signaling Pathway. *J. Steroid Biochem. Mol. Biol.* **2009,** *116* (1–2), 50–55.

2. Banerjee, S.; Li, Y.; Wang, Z.; Sarkar, F. H. Multi-Targeted Therapy of Cancer by Genistein. *Cancer Lett.* **2008,** *269* (2), 226–242.

3. Davis, J. N.; Kucuk, O.; Djuric, Z.; Sarkar, F. H. Soy Isoflavone Supplementation in Healthy Men Prevents NF-Kappa B Activation by TNF-Alpha in Blood Lymphocytes. *Free Radical Bio. Med.* **2001,** *30* (11), 1293–1302.

4. Dell' Agli, M.; Fagnani, R.; Galli, G. V. Olive Oil Phenols Modulate the Expression of Metalloproteinase 9 in THP-1 Cells by Acting on Nuclear Factor-Kappa B Signaling. *J. Agr. Food Chem.* **2010,** *58* (4), 2246–2252.

5. Droge, W. Free Radicals in the Physiological Control of Cell Function. *Physiol. Rev.* **2002,** *82* (1), 47–95.

6. Grynkiewicz, G.; Slifirski, P. Curcumin and Curcuminoids in Quest for Medicinal Status. *Acta Biochim. Pol.* **2012,** *59* (2), 201–212.

7. Guix, F. X. The Physiology and Pathophysiology of Nitric Oxide in the Brain. *Prog. Neurobiol.* **2005,** *76* (2), 126–152.

8. Halliwell, B.; Gutteridge, J. Antioxidant Defense: Endogenous and Diet Derived. *J. Free Radic. Biol. Med.* **2007,** *79,* 186.

9. Hastak, K.; Agarwal, M. K.; Mukhtar, H.; Agarwal, M.L. Ablation of Either p21 or Bax Prevents p53-Dependent Apoptosis Induced by Green Tea Polyphenol Epigallocatechin-3-Gallate. *FASEB J.* **2005,** *19* (7), 789–791.

10. Jaeger, B. N.; Parylak, S. L.; Gage, F. H. Mechanisms of Dietary Flavonoid Action in Neuronal Function and Neuroinflammation. *Mol. Aspects Med.* **2018,** *61,* 50–62.

11. Joseph, J. A.; Shukitt-Hale, B.; Brewer, G. J.; Weikel, K. A.; Kalt, W.; Fisher, D. R. Differential Protection Among Fractionated Blueberry Polyphenolic Families Against

DA-, Aβ_{42}- and LPS-Induced Decrements in Ca^{2+} Buffering in Primary Hippocampal Cells. *J. Agr. Food Chem.* **2010,** *58* (14), 8196–8204.

12. Jurenka, J. S. Anti-Inflammatory Properties of Curcumin, a Major Constituent of Curcuma Longa: A Review of Preclinical and Clinical Research. *Altern. Med. Rev.* **2009,** *14* (2), 141–153.

13. Kim, T. I.; Lee, Y. K.; Park, S. G. L-Theanine, An Amino Acid in Green Tea, Attenuates Beta-Amyloid-Induced Cognitive Dysfunction and Neurotoxicity: Reduction in Oxidative Damage and Inactivation of ERK/p38 Kinase and NF-Kappa B Pathways. *Free Rad. Biol. Med.* **2009,** *47* (11), 1601–1610.

14. Kunnumakkara, A. B.; Diagaradjane, P.; Guha, S. Curcumin Sensitizes Human Colorectal Cancer Xenografts in Nude Mice to Gamma-Radiation by Targeting Nuclear Factor-Kappa B-Regulated Gene Products. *Clin. Cancer Res.* **2008,** *14* (7), 2128–2136.

15. Kuriyama, S.; Hozawa, A.; Ohmori, K.; Shimazu, T. Green Tea Consumption and Cognitive Function: A Cross-Sectional Study from the Tsurugaya Project 1. *Am. J. Clin. Nutr.* **2006,** *83* (2), 355–361.

16. Lee, J. W.; Lee, Y. K.; Ban, J. O.; Ha, T. Y.; Yun, Y. P. Green Tea (-)- Epigallocatechin-3-Gallate Inhibits Beta-Amyloid-Induced Cognitive Dysfunction through Modification of Secretase Activity via Inhibition of ERK and NF-kappa B Pathways in Mice. *J. Nutr.* **2009,** *139*, 1987–1993.

17. Li, Y.; Ahmed, F.; Ali, S.; Philip, P. A.; Kucuk, O.; Sarkar, F. H. Inactivation of Nuclear Factor Kappa B by Soy Isoflavone Genistein Contributes to Increased Apoptosis Induced by Chemotherapeutic Agents in Human Cancer Cells. *Cancer Res.* **2005,** *65* (15), 6934–6942.

18. Litescu, S. C.; Eremia, S.; Radu, G. L. Methods for Determination of Antioxidant Capacity in Food and Raw Materials. *Adv. Exp. Med. Biol.* **2010,** *698*, 241–249.

19. Loke, W. M.; Proudfoot, J. M.; Hodgson, J. M. Dietary polyphenols attenuate atherosclerosis in apolipoprotein E-Knockout Mice by Alleviating Inflammation and Endothelial Dysfunction. *Arterioscler. Thromb. Vasc. Biol.* **2010,** *30* (4), 749–757.

20. Madhyastha, R.; Madhyastha, H.; Nakajima, Y.; Omura, S.; Maruyama, M. Curcumin Facilitates Fibrinolysis and Cellular Migration during Wound Healing by Modulating Urokinase Plasminogen Activator Expression. *Pathophysiol. Haemo. Thromb.* **2010,** *37* (2–4), 59–66.

21. Mahmoud, A. M.; Yang, W.; Bosland, M. C. Soy Isoflavones and Prostate Cancer: A Review of Molecular Mechanisms. *J. Steroid Biochem. Mol. Biol.* **2014,** *140*, 116–132.

22. Manach, C.; Scalbert, A.; Morand, C.; Remesy, C.; Jimenez, L. Polyphenols: Food Sources and Bioavailability. *Am. J. Clin. Nutr.* **2004,** *79* (5), 727–747.

23. Mancuso, C.; Scapagnini, G.; Curro, D.; Giuffrida Stella A. M.; De Marco, C.; Butterfield, D. A.; Calabrese, V. Mitochondrial Dysfunction, Free Radical Generation and Cellular Stress Response in Neurodegenerative Disorders. *Front. Biosci.: J. Virt. Lib.* **2007,** *12*, 1107–1123.

24. Natsume, M.; Baba, S. Suppressive Effects of Cacao Polyphenols on the Development of Atherosclerosis in Apolipoprotein E-Deficient Mice. *Subcell. Biochem.* **2014,** *77*, 189–198.

25. Nishikawa, T.; Nakajima, T.; Moriguchi, M.; Jo, M. A Green Tea Polyphenol, Epigalocatechin-3-Gallate, Induces Apoptosis of Human Hepatocellular Carcinoma, Possibly Through Inhibition of Bcl-2 Family Proteins. *J. Hepatol.* **2006,** *44* (6), 1074–1082.

26. Paduch, R.; Kandefer-Szerszen, M.; Piersiak, T. The Importance of Release of Proinflammatory Cytokines, ROS, and NO in Different Stages of Colon Carcinoma Growth and Metastasis after Treatment with Cytotoxic Drugs. *Oncol. Res.* **2010,** *18* (9), 419–436.

27. Pandey, K. B.; Rizvi, S. I. Plant Polyphenols as Dietary Antioxidants in Human Health and Disease. *Oxid. Med. Cell. Longev.* **2009,** *2* (5), 270–278.

28. Pengjam, Y.; Madhyastha, H.; Madhyastha, R. Glycoside Aloin Induces Osteogenic Initiation of MC3T3-E1 Cells: Involvement of MAPK Mediated Wnt and Bmp Signaling. *Biomol. Ther.* **2016,** *24* (2), 123–131.

29. Pengjam, Y.; Madhyastha, H.; Madhyastha, R. M -kB Pathway Inhibition by Anthrocyclic Glycoside Aloin Is Key Event in Preventing Osteoclastogenesis in RAW 264.7 Cells. *Phytomedicine* **2016,** *23*, 417–428.

30. Pianetti, S.; Guo, S.; Kavanagh, K. T.; Sonenshein, G. E. Green Tea Polyphenol Epigallocatechin-3 Gallate Inhibits Her-2/neu Signaling, Proliferation, and Transformed Phenotype of Breast Cancer Cells. *Cancer Res.* **2002,** *62* (3), 652–655.

31. Quideau, S.; Deffieux, D.; Douat-Casassus, C.; Pouységu, L. Plant Polyphenols: Chemical Properties, Biological Activities, and Synthesis. *Angew. Chem.* **2011,** *50* (3), 586–621.

32. Rossi, L.; Mazzitelli, S.; Arciello, M.; Capo, C. R.; Rotilio, G. Benefits from Dietary Polyphenols for Brain Aging and Alzheimer's Disease. *Neurochem. Res.* **2008,** *33* (12), 2390–2400.

33. Sah, J.; Balasubramanian, S.; Eckert, R. L.; Rorke, E. A. Epigallocatechin-3-Gallate Inhibits Epidermal Growth Factor Receptor Signaling Pathway. Evidence for Direct Inhibition of ERK1/2 and AKT Kinases. *J. Biol. Chem.* **2004,** *279* (13), 12755–12762.

34. Sahin, K.; Tuzcu, M.; Sahin, N.; Akdemir, F. Inhibitory Effects of Combination of Lycopene and Genistein on 7,12-Dimethyl benz(a)anthracene-Induced Breast Cancer in Rats. *Nutr. Cancer* **2011,** *63* (8), 1279–1286.

35. Scoditti, E.; Calabriso, N.; Massaro, M. Mediterranean Diet Polyphenols Reduce Inflammatory Angiogenesis through MMP-9 and COX-2 Inhibition in Human Vascular Endothelial Cells: A Potentially Protective Mechanism in Atherosclerotic Vascular Disease and Cancer. *Arch. Biochem. Biophy.* **2012,** *527* (2), 81–89.

36. Sgarbossa, A.; Giacomazza, D.; di Carlo, M. Ferulic Acid: A Hope for Alzheimer's Disease Therapy from Plants. *Nutrients* **2015,** *7* (7), 5764–5782.

37. Sies, H. Oxidative Stress: Oxidants and Antioxidants. *Exp. Physiol.* **1997,** *82* (2), 291–295.

38. Singh, B. N.; Shankar, S.; Srivastava, R. K. Green Tea Catechin Epigallocatechin-3-Gallate (EGCG): Mechanisms, Perspectives and Clinical Applications. *Biochem. Pharmacol.* **2011,** *82* (12), 1807–1821.

39. Spencer, J. P. Flavonoids and Brain Health: Multiple Effects Unpinned by Common Mechanisms. *Genes Nutr.* **2009,** *4* (4), 243–250.

40. Spencer, J. P. The Impact of Flavonoids on Memory: Physiological and Molecular Considerations. *Chem. Soc. Rev.* **2009,** *38* (4), 1152-1161.

41. Tangney, C. C.; Rasmussen, H. E. Polyphenols, Inflammation, and Cardiovascular Disease. *Curr. Atheroscler. Rep.* **2013,** *15* (5), 324.

42. Thangapazham, R. L.; Singh, A. K.; Sharma, A.; Warren, J. Green Tea Polyphenols and Its Constituent Epigallocatechin Gallate Inhibits Proliferation of Human Breast Cancer Cells In Vitro and In Vivo. *Cancer Lett.* **2007,** *245* (1–2), 232–241.

43. Uchino, R.; Madhyastha, R.; Madhyastha, H.; Dhungana, S. NFkB-Dependent Regulation of Urokinase Plasminogen Activator by Proanthocyanidin-Rich Grape Seed

Extract: Effect on Invasion by Prostate Cancer Cells. *Blood Coagul. Fibrin.: Int. J. Haemo. Thromb.* **2010**, *21*, 528–533.

44. Wang, Z.; Zhang, Y.; Li, Y.; Banerjee, S.; Laio, J.; Sarkar, F. H. Down-Regulation of Notch-1 Contributes to Cell Growth Inhibition and Apoptosis in Pancreatic Cancer Cells. *Mol. Cancer Ther.* **2006**, *5* (3), 483–493.

45. Weir, N. M.; Selvendiran, K.; Kutala, V. K.; Tong, L. Curcumin Induces G2/M Arrest and Apoptosis in Cisplatin-Resistant Human Ovarian Cancer Cells by Modulating Akt and p38 MAPK. *Cancer Biol. Ther.* **2007**, *6* (2), 178–184.

46. Widmer, R. J.; Freund, M. A.; Flammer, A. J.; Sexton, J. Beneficial Effects of Polyphenol-Rich Olive Oil in Patients with Early Atherosclerosis. *Eur. J. Nutr.* **2013**, *52* (3), 1223–1231.

47. Williams, C. M.; El Mohsen, M. A.; Vauzour, D. Blueberry-Induced Changes in Spatial Working Memory Correlate with Changes in Hippocampal CREB Phosphorylation and Brain-Derived Neurotrophic Factor (BDNF) Levels. *Free Rad. Biol. Med.* **2008**, *45* (3), 295–305.

48. Woo, J. H.; Kim, Y. H.; Choi, Y. J.; Kim, D. G. Molecular Mechanisms of Curcumin-Induced Cytotoxicity: Induction of Apoptosis through Generation of Reactive Oxygen Species, Down-Regulation of Bcl-XL and IAP, the Release of Cytochrome C and Inhibition of Akt. *Carcinogenesis* **2003**, *24* (7), 1199–1208.

49. Yallapu, M. M.; Jaggi, M.; Chauhan, S. C. Curcumin Nanomedicine: A Road to Cancer Therapeutics. *Curr. Pharm. Des.* **2013**, *19*(11), 1994–2010.

50. Yan, J. J.; Cho, J. Y.; Kim, H. S.; Kim, K. L. Protection Against β-amyloid Peptide Toxicity In Vivo with Long-Term Administration of Ferulic Acid. *Br. J. Pharmacol.* **2001**, *133* (1), 89–96.

51. Zhang, J.; Wang, X.; Vikash, V.; Ye, Q.; Wu, D.; Liu, Y.; Dong, W. ROS and ROS-Mediated Cellular Signaling. *Oxid. Med. Cell. Longev.* **2016**, *2016*, 4350965.

52. Zhou, Y.; Zheng, J.; Li, Y.; Xu, D. P.; Li, S.; Chen, Y. M.; Li, H. B. Natural Polyphenols for Prevention and Treatment of Cancer. *Nutrients* **2016**, *8* (8), 515.

CHAPTER 15

HYALURONIDASE AND GELATINASE (MMP-2, MMP-9) INHIBITOR PLANTS

C. DONMEZ, G. D. DURBILMEZ, H. EL-SEEDI, and
U. KOCA-CALISKAN

ABSTRACT

This chapter emphasizes enzyme inhibitory activities of natural plants and their compounds, specifically hyaluronidase and the gelatinase (matrix metalloproteinase (MMP)-2 and MMP-9) inhibitor medicinal plants and their contents. Particularly, different herbal plants have hyaluronidase enzyme inhibitor potentials. Evaluation of such plants in Rosaceae, Asteraceae, Lamiaceae, and Zingiberaceae families has shown that the extracts, tannins, and polyphenols were identified/isolated as enzyme inhibitors. Gelatinases, inhibitions of MMP-2 and MMP-9, are the most common in Asteraceae, Fabaceae, Lamiaceae, and Rosaceae families. Mostly, saponin and flavonoid derivatives have important role in the inhibition of MMPs activities.

15.1 INTRODUCTION

Hyaluronic acid (HA) and gelatin are natural substances of not only skin but also other organs. Both of these play different roles in various parts of the body, more specifically in aging and skin metabolism, such as skin hydration, balance, vitality, softness, and elasticity.[15] HA is a mucopolysaccharide molecule, which has water holding feature, whereas gelatin is a peptide molecule that has an anti-aging effect by providing the passage of collagen into tissues.

Although both have substantial effects, level of HA and gelatin in the skin decreases due to aging, improper diet, stress, and various external factors, such as sunlight, wind, irritation, injury, and cigarette smoke. The degradation of HA and gelatin is undesirable in terms of maintaining vitality and flexibility of the skin. "Replacement therapy", "increasing the synthesis of the related enzymes", and "inhibiting their disintegration" on the skin are some of the solutions to keep the level of the enzymes at desirable levels. "Hyaluronidase" is an enzyme that catalyzes the degradation of HA. "Gelatinase A and B" (also known as "matrix metalloproteinase (MMP)-2 and MMP-9") are members of the matrix metalloproteinase enzyme family to catalyze the degradation of gelatin.

Therefore, the inhibition of degrading enzymes is crucial. Major hyaluronidase and gelatinase inhibition can be performed with various synthetic chemical substances (e.g. MMP inhibitors; batimastat, marimastat, prinomastat, hydroxyproline and hydroxamic acid derivatives, hydantoin–hyaluronidase inhibitors; polycondensed diphenyl-methane and triphenyl methane derivatives). Inevitable side effects and safety risks of chemicals led researchers to search for natural/herbal enzyme inhibitors. Recent studies have shown that various plants have inhibitory effects on hyaluronidase and gelatinase enzymes. Plants in traditional folk medicine are suitable candidates for research.

Many popular cosmetic products often used to prevent skin aging and skin wrinkles contain herbal ingredients, in which some of these are enzyme inhibitors from plant extracts. The scope of this chapter is to explain medicinal plants and their compounds, which have MMPs and hyaluronidase inhibitory activities.

This chapter emphasizes on the hyaluronidase and gelatinase (MMP-2 and MMP-9) inhibitors in medicinal plants/phytochemicals.

15.2 ENZYMES AND INHIBITORS

15.2.1 HYALURONIDASE

HA, the major glycosaminoglycan component of the extracellular matrix, depends on hyaluronidase enzyme activity. Hyaluronoglucosidase/Hyaluro-noglucosaminidase destroys HA by the "(1,4)-linkages between N-acetyl-β-D-glucosamine" and "D-glucouronate" residues in hyaluronate–hyaluronan, which is a polymer in joint and tissue matrices.[14,62,64] Hyaluronidase includes five different human enzymes. Three of these are measurable activities for endolytic hyaluronan hydrolysis. One of these plays an important role on chondroitin sulfate, and potential of fifth enzyme has not yet been defined. Hyaluronidase-1 increases in bladder, colorectal, breast, and ovarian cancers. Hyaluronidase-1 and 3 showed overexpression in lung cancer.[41] Hyaluroni-dase-2 was identified in neck squamous cell carcinoma.[25] For the last 60 years, hyaluronidase has been used in ophthalmic surgery, dermatosurgery, and in other surgical fields. In cosmetic formulations, hyaluronic acid is utilized too much. Nearly, 50% of all hyaluronan in the body is located in skin tissues.[8]

15.2.1.1 PLANT-BASED HYALURONIDASE INHIBITORS

Recently, herbal materials are preferred mostly in cosmetic and cosmeceutical products for enhancing hyaluronic acid on skin. The principal purpose is to prevent the breakdown of hyaluronic acid thus increasing the hyaluronic acid/hyaluronan.

Tannin-rich plants (such as *Potentilla erecta* (L.) Raeusch, *Rubus fruticosus* L., *Rubus idaeus* L., and *Quercus robur* L.) have an important anti-hyaluronidase effect.[57] It has been shown that "Shikimic acid" isolated from *Alnus glutinosa* (L.) Gaertn. subsp. *glutinosa* and "ascorbic acid" from *Cucumis sativus* L. have inhibitory influence on hyaluronidase.[2,15] On the other hand, the extracts from *Ononis* and *Solidaginis* species inhibit the activity of hyaluronidase enzymes.[53] Table 15.1 indicates selected plant-based inhibitory hyaluronidase enzyme.

15.2.2 GELATINASES/METALLOPROTEINASES

Gelatinase is a proteolytic enzyme that catalyzes the breakdown of gelatin. Gelatinase includes MMPs that are found in human, vertebrate, and nonvertebrate. [66] MMPs, belonging to the "Metzincins" superfamily, are calcium-dependent. These include MMP-2 (gelatinase-A) and MMP-9 (gelatinase-B). Their basic activity is to hydrolyze the gelatin to amino acid. Further, it reduces the extracellular matrix components (ECM) like elastin, fibronectin, and collagens (types IV, V, VII, and X) in helical domains.[35] While activated MMP-2 may connect to integrin $\alpha v\beta 3$ on the uppermost layer of angiogenic endothelial cells and spreading cancer cells, MMP-9 binds to type IV collagen $\alpha 2$ chains on the surface of various cell types especially in skin cancers.[7,51] Moreover, angiogenesis and tumor cell invasion are supported via the localizations of MMPs to the cell surfaces. In malignant cancers, MMPs are activated and their levels are increased. Therefore, inhibition of these enzymes is required for treatments of many cancers.[66] MMPs are involved in numerous biological and physiological processes, such as embryonic development, inflammatory response, tissue morphogenesis, wound repairing, bone remodeling, autoimmunity, and cell growth.[45,50,52]

TABLE 15.1 Hyaluronidase Inhibitory Plants.

Latin name	Family	Effective extract or biocompound(s)	References
Alnus glutinosa (L.) Gaertn. subsp. *Glutinosa*	Betulaceae	Shikimic acid	[2]
Alpinia katsumadai Hayata	Zingiberaceae	Chalcone; flavonoids; sesquiterpenoids	[15]
Alpinia zerumbet (Pers.) B.L. Burtt & R.M.Sm.	Zingiberaceae	Dihydro-5,6-dehydrokawain 8 (17),12-labdadiene-15,16-dial; 5,6-dehydrokawain	[13]
Amaranthus viridis L.	Amaranthaceae	Extract	[63]
Biophytum sensitivum (L.) DC.	Oxalidaceae	Amentoflavone; cupressuflavon; isoorientin	[15]
Chamaerhodos altaica Bunge	Rosaceae	Extract; Potentillin; eugenin; agrimoniin; 1,2,6-tri-*O*-galloyl-β-D-glucopyranoside; 1,2,3,4,6-penta-*O*-galloyl-β-D-glucopyranoside; (2*R*,3*S*)-3,4-dihydro-2-(3,4-dihydroxyphenyl)-2*H*-chromene-3,5,7-triol	[64]
Chamaerhodos erecta (L.) Bunge	Rosaceae	Extract; Potentillin; eugenin; agrimoniin; 1,2,6-tri-*O*-galloyl-β-D-glucopyranoside; 1,2,3,4,6-penta-*O*-galloyl-β-D-glucopyranoside; (2*R*,3*S*)-3,4-dihydro-2-(3,4-dihydroxyphenyl)- 2*H*-chromene-3,5,7-triol	[64]
Cinnamomum cassia (L.) J. Presl	Lauraceae	Phenolics; isoeugenol; cinnamic aldehyde; coumarins	[15]
Cucumis sativus L.	Cucurbitaceae	Ascorbic acid	[15]
Curcuma longa L.	Zingiberaceae	Curcuminoids; curcumin; demethoxycurcumin; bisdimethoxycurcumin	[15]
Dracocephalum foetidum Bunge	Lamiaceae	Acacetin-7-*O*-(3,6-*O*-dimalonyl)-β-D-glucopyranoside; acacetin-7-*O*-(3-*O*-malonyl)-β-D-glucuronopyranoside; acacetin-7-*O*-β-D-glucuronide; apigenin 7-*O*-(6-malonyl-β-D-glucoside); apigenin 7-*O*-β-glucuronide; rosmarinic acid; luteolin-7-*O*-β-D-glucuronide	[64]

TABLE 15.1 (Continued)

Latin name	Family	Effective extract or biocompound(s)	References
Dryopteris crassirhizoma Nakai	Dryopteridaceae	Dryopteroside; [+]catechin-6-C-β-D-glucopyranoside	[15]
Equisetum arvense L.	Equisetaceae	Extract; Phenolic compounds; flavonoids	[53, 64]
Euphorbia hirta L.	Euphorbiaceae	Quercetin-3-O-α-rhamnoside	[16]
Garcinia indica (Thouars) Choisy	Clusiaceae	Garcinol; isogarcinol; cyanidin; coumarins	[15]
Gaultheria procumbens L.	Ericaceae	Extract	[43]
Geranium pratense L.	Geraniaceae	Extract; Tannins; polyphenols	[57]
Geranium robertianum L.	Geraniaceae	Extract; Tannins; polyphenols	[57]
Geum urbanum L:	Rosaceae		[57]
Glycyrrhiza glabra L.	Fabaceae	Extract; Glycyrrhizin	[15]
Impatiens parviflora DC	Balsaminaceae	Phenols; flavonoids; coumarins; triterpenes; saponins; carotenoids; sterols; lipids	[15, 17]
Leucas aspera Link	Lamiaceae	Alkaloids; phenolic compounds; ursolic acid	[15]
Lythrum salicaria L.	Lythraceae	Extract; Tannins; polyphenols	[57]
Lythrum salicaria L.	Lythraceae	Extract; Salicarinins A, B and C; vescalagin; castalagin	[64]
Malva sylvestris L.	Malvaceae	Extract	[53]
Myristica fragrans Houtt.	Myristicaceae	Volatile oils, myristicin; myristic, EC; cyanidin	[15]
Oenothera biennis L.	Onagraceae	Extract	[64]
Ononis L. sp.	Fabaceae	Extract	[53]
Panax japonicus (T.Nees) C. A. Mey.	Araliaceae	Saponins; catechin	[70]
Periploca forrestii Schltr	Apocynaceae	Flavonoids	[33]

TABLE 15.1 (Continued)

Latin name	Family	Effective extract or biocompound(s)	References
Potentilla erecta (L.) Raeusch.	Rosaceae	Extract Tannins; polyphenols	[57]
Prunus salicina Lindl.	Rosaceae	Saponins; catechin	[70]
Quercus robur L.	Fagaceae	Extract; Tannins; polyphenols	[57]
Rubus fruticosus L.	Rosaceae	Extract; Flavonoids (flavonols, flavonones); catechin, anthocyanidins; tannins, polyphenols	[40, 57]
Rubus idaeus L.	Rosaceae	Extract; Tannins; polyphenols	[57]
Solidaginis L. sp.	Asteraceae	Extract;	[53]
Tagetes erecta L.	Asteraceae	Syringic acid; provitamin A "carotenoids"; β-amyrin	[15]
Terminalia chebula Retz.	Combretaceae	Hydrolysable tannins (gallotannins, gallic acid, casuarinin, chebulinic acid, chebulanin, corilagin, terchebulin, neochebulinic acid); phenolics (chebulic acid, ellagic acid)	[15]
Tessaria absinthioides (Hook. & Arn.) DC.	Asteraceae	Extract	[71]
Thymus zygis L.	Lamiaceae	Thymol	[9]

15.2.2.1   PLANT-BASED METALLOPROTEINASES INHIBITORS

Plurality of synthetic MMP inhibitors has been described in the past.[44] Collagen peptidomimetic, biphosphonate inhibitors, and tetracycline derivatives, which have passed phase III trials, are in this group.[36] Since anticipated potential from synthetic inhibitors is not seen, it has been directed to discover new molecules from natural sources.

Mukherjee et al. mention 70 different plants that inhibit gelatinase enzymes. Both plants and their constituents have effective inhibitory activity.

Several studies about phytochemical MMPs inhibitors have been published. "A prenylated chalcone derived (xanthohumol) from *Humulus lupulus* L.", "Flavonoid glycosides (isorhamnetin 3-*O* {cyrillic}-*β*-D-glucoside and quercetin 3-*O*-*β*-D-glucoside) from *Salicornia herbacea*", "Rhizoma Paridis saponins/Resveratrol/saponin derived from *Platycodon grandiflorum*" have important inhibitory role in metastasis of cancer by restricting the expressions of MMP-2 and -9.[24,26,38,48] It is reported that catechin, gallic and protocatechuic acid, epicatechin (EC), epi-gallocatechin gallate (EGCG), caffeic acid, and rutin have chemoprotective effects due to the inhibition of MMP-2 and -9 protein expressions.[10,68] "4-Nerolidylcatechol isolated from *Pothomorphe umbellata* (L.) Miq." reduces expression of MMPs that increases with ultraviolet radiation.[6,47] "Sinomenine isolated from *Sinomenium acutum*" shows inhibitory effects in inflammatory autoimmune diseases.[54]

It is reported that antiphlogistic *Achillea millefolium* L., anti-cancer *Salvia miltiorrhiza* Bunge, *Elaeagnus glabra* (Thunb.), *Magnolia officinalis*, and *Selaginella tamariscina* have suppressed both MMP-2 and -9.[5,28,32, 65,72] *Pothomorphe umbellata* can inhibit pro-MMP-2, MMP-2 and MMP-9.[3,60] Moreover, phlorotannins isolated from brown algae (known as *Ecklonia cava*) have shown inhibitory functions of MMP-2 and -9 on human skin fibroblasts and HT-1080 cells.[23]

Extract of *Cynara cardunculus* L. var. *scolymu*s (L.) Benth inhibits MMP-2 enzyme activity.[44] "Clinopodic acid-C isolated from *Clinopodium chinense* var. *parviflorum*", "cyanidin glycosides from *Morus alba* L.", and "obovatal from *Magnolia obovata*" have shown their MMP-2 down-regulatory activities.[11,29,49]

Many studies have shown that flavonoid and saponin derivatives have potentially MMP-9 inhibitory effects.[1,20,30,37,73,74] Decursin, a coumarin isolated from *Angelicae gigas*, has diminished expression of MMP-9 by preventing phosphorylation.[21]

Some plants related to MMP-2 and -9 enzyme inhibitions are listed in Table 15.2.

TABLE 15.2 MMP-2 and -9 Inhibitory Plants.

Latin name	Family	Effective extract or biocompound	Inhibitor	References
Achillea millefolium L.	Asteraceae	Extract	MMP-2 and -9	[5]
Angelica gigas Nakai	Apiaceae	Decursin	MMP-9	[21]
Artemisia capillaries Thunberg	Asteraceae	Capillarisin	MMP-9	[30]
Calendula officinalis L.	Asteraceae	Triterpenoids; flavonoids (patuletin and Patulitrin); saponins	MMP-2 and -9	[15]
Camellia sinensis (L.) Kuntze	Theaceae	Epi-gallocatechin-3-*O*-gallate		[68]
Cassia fistula L.	Fabaceae	Extract	MMP-2	[34]
Citrus L. *sp.*	Rutaceae	Nobiletin	MMP-9	[19]
Cleistanthus collinus Benth. & Hook.f.	Phyllanthaceae	Cleistanthin A		[42]
Cleistocalyx nervosum (DC.) Kosterm. var. *paniala*	Myrtaceae	Extract; Phenolic compounds	MMP-2	[39]
Clinacanthus nutans Lindau	Acanthaceae	Vitexin; orientin; isovitexin; isoorientin; lupeol; β-sitosterol; clinacoside A-C; cycloclinacoside A1;shaftoside	MMP-2 and MMP-9	[58]
Clinopodium chinense var. parviflorum (Kudô) H. Hara	Lamiaceae	Clinopodic acid C	MMP-2	[49]
Combretum hartmannianum Schweinf.	Combretaceae	Terchebulin	MMP-9	[45]
Crataegus pinnatifida Bunge	Rosaceae	Epicatechin-(4- β -6)- epicatechin-(4- β -8)- epicatechin	MMP-2 and -9	1001 [46]
Cucumis melo L.	Cucurbitaceae	Extract		[60]
Curcuma longa L.	Zingiberaceae	Curcuminoids; curcumin; demethoxycurcumin; bisdimethoxy curcumin		[15, 76]
Cynara cardunculus L.	Asteraceae	Extract Luteolin; apigenin; caffeic acid	MMP-9	[4]
Cynara cardunculus L. var. *scolymus* (L.) Benth.	Asteraceae	Extract	MMP-2	[44]

TABLE 15.2   (Continued)

Latin name	Family	Effective extract or biocompound	Inhibitor	References
Davallia bilabiata Hosok.	Davalliaceae	Epicatechin 3-O-β-D-allopyranoside	MMP-2	[36]
Elaeagnus glabra Thunb.	Elaeagnaceae	Extract	MMP-2 and -9	[32]
Euonymus alatus (Thunb.) Sieb.	Celastraceae	Caffeic acid (3,4-dihydroxycinnamic acid); 5-caffeoylquinic acid.	MMP-9	[55, 68]
Geum japonicum Thunb.	Rosaceae	3,4,5-Trihydroxy-benzaldehyde		[67]
Ginkgo biloba L.	Ginkgoaceae	Extract		[37]
Glycyrrhiza glabra L.	Fabaceae	Lipopolysaccharide; D-galactosamine		[1]
Humulus lupulus L.	Cannabinaceae	Xanthohumol	MMP-2 and -9	[22]
Kaemphferia pandurata Roxb.	Zingiberaceae	Panduratin A	MMP-9	[73]
Machilus thunbergii Siebold & Zucc.	Lauraceae	Meso-dihydroguaiaretic acid		[75]
Magnolia obovata Thunb.	Magnoliaceae	Lignans (obovatal, obovatol)	MMP-2	[29]
Magnolia officinalis Rehder & E.H.Wilson	Magnoliaceae	Magnolol	MMP-2 and -9	[20, 28]
Metasequoia glyptostroboides Hu & W.C.Cheng	Cupressaceae	Isoginkgetin	MMP-9	[74]
Morus alba L.	Moraceae	Catechin, protocatechuic acid, gallic acid, EGCG, EC, caffeic acid, cyanidin-3-glucoside, rutin, cyanidin-3-rutinoside	MMP-2 and -9	[10, 11]
Paris polyphylla Smith	Melanthiaceae	Saponins		[38]
Phlebodium aureum (L.) J.Sm.	Polypodiaceae	Extract		[56]
Platycodon grandiflorum A.DC.	Campanulaceae	Saponins		[26]
Pothomorphe umbellata (L.) Miq.	Piperaceae	Steroids; sesquiterpenes; volatile oils; 4-nerolidylcatechol,	Pro-MMP-2, MMP-2 and MMP-9	[3, 6, 47, 61]
Pterocarpus santalinus L.f.	Fabaceae	Terpenoids; pigments-santalin; erthyrodiol	MMP-2	[15]

TABLE 15.2 *(Continued)*

Latin name	Family	Effective extract or biocompound	Inhibitor	References
Rhodiola rosea L.	Crassulaceae	1,2,3,6-tetra-*O*-galloyl-4-*O*-*p*-hydroxybenzoyl-β-D-glucopyranoside; 1,2,3,6-tetra-*O*-galloyl-β-D-glucopyranonoside		[31]
Rhodomyrtus tomentosa (Aiton) Hassk.	Myrtaceae	Rhodomyrtone	MMP-2 and -9	[69]
Rosmarinus officinalis L.	Lamiaceae	Extract Rosmarinic acid	MMP-2	[59]
Rubus idaeus L.	Rosaceae	Extract		[18]
Salicornia herbacea L.	Cheno-podiaceae	Flavonoids (quercetin 3-*O*-β-D-glucoside; isorhammetin 3-*O* {cyrillic}-β-D-glucoside)	MMP-2 and -9	[24]
Salvia miltiorrhiza Bunge	Lamiaceae	Extract		[65]
Selaginella tamariscina (P.Beauv.) Spring	Selagi-nellaceae	Extract		[72]
Sinomenium acutum (Thunb.) Rehder & E.H.Wilson	Meni-spermaceae	Sinomenine		[54]
Terminalia chebula Retz.	Combretaceae	Hydrolysable tannins (casuarinin, chebulinic acid, neochebulinic acid, gallotannins, chebulanin, terchebulin, corilagin, and gallic acid); phenolics (chebulic acid, ellagic acid)	MMP-2	[15]
Theobroma cacao L.	Sterculiaceae	Procyanidin B2	Pro-MMP-2	[27]
Trillium tschonoskii Maxim.	Trilliaceae	Extract Saponins	MMP-2 and -9	[12]

15.3 SUMMARY

Hyaluronic acid and gelatin are natural substances of skin. MMPs are considered as cancer indicators. Hyaluronidase enzyme is usually responsible for maintaining vitality and flexibility of the skin. The degradation of hyaluronic acid and gelatin is undesirable for our body. Therefore, the inhibition of these enzymes is important. Inhibition of hyaluronidase and gelatinase can be performed with various synthetic chemical substances. Side effects and safety risk of chemicals have led people to prefer natural/herbal products. In recent years, many studies show that various plants have an inhibitory effect on hyaluronidase and gelatinase. This chapter highlights the gelatinase (MMP-2 and MMP-9) and hyaluronidase inhibitors medicinal plants/phytochemicals. Many studies have shown that flavonoid, saponin, tannin, and coumarin derivatives have potential inhibitory effects.

KEYWORDS

- gelatinase
- herbal enzyme inhibitors
- hyaluronidase
- matrix metalloproteinase
- MMP-2
- MMP-9

REFERENCES

1. Abe, K.; Ikeda, T.; Wake, K.; Sato, T.; Sato, T.; Inoue, H. Glycyrrhizin Prevents Of Lipopolysaccharide/D-Galactosamine-Induced Liver Injury Through Down-Regulation Of Matrix Metalloproteinase-9 In Mice. *J. Pharm. Pharmacol.* **2008,** *60* (1), 91–97.
2. Altınyay, Ç.; Süntar, I.; Altun, L.; Keleş, H.; Akkol, E. K. Phytochemical And Biological Studies On *Alnus Glutinosa* Subsp. *Glutinosa*, *A. Orientalis* Var. *Orientalis* And *A. Orientalis* Var. *Pubescens* Leaves. *J. Ethnopharmacol.* **2016,** *192*, 148–160.
3. Barros, L. F. M.; Barros, P. S. M.; Ropke, C. D. Dose-Dependent In Vitro Inhibition Of Rabbit Corneal Matrix Metalloproteinases By An Extract Of *Pothomorphe Umbellata* After Alkali Injury. *Braz. J. Med. Biol. Res.* **2007,** *40* (8), 1129–1132.

4. Bellosta, S.; Bogani, P.; Canavesi, M.; Galli, C.; Visioli, F. Mediterranean Diet and Cardioprotection: Wild Artichoke Inhibits Metalloproteinase 9. *Mol. Nutr. Food Res.* **2008,** *52* (10), 1147–1152.

5. Benedek, B.; Kopp, B.; Melzig, M. F. *Achillea millefolium* L. s.l. – Is The Anti-Inflammatory Activity Mediated By Protease Inhibition? *J. Ethnopharmacol.* **2007,** *113* (2), 312–317.

6. Brohem, C. A.; Sawada, T. C. H.; Massaro, R. R. Apoptosis Induction By 4-Nerolidylcatechol In Melanoma Cell Lines. *Toxicol. In Vitro* **2009,** *23* (1), 111–119.

7. Brooks, P. C.; Strömblad, S.; Sanders, L. C. Localization of Matrix Metalloproteinase MMP-2 To The Surface Of Invasive Cells By Interaction With Integrin αvβ3. *Cell* **1996,** *85* (5), 683–693.

8. Buhren, B. A.; Schrumpf, H.; Hoff, N. P.; Bölke, E.; Hilton, S.; Gerber, P. A. Hyaluronidase: From Clinical Applications To Molecular And Cellular Mechanisms. *Eur. J. Med. Res.* **2016,** *21* (1), 5–9.

9. Carrasco, A.; Tomas, V.; Tudela, J.; Miguel, M. G. Comparative Study Of GC-MS Characterization, Antioxidant Activity And Hyaluronidase Inhibition Of Different Species Of Lavandula And Thymus Essential Oils. *Flavour Frag. J.* **2016,** *31* (1), 57–69.

10. Chan, K. C.; Ho, H. H.; Huang, C. N.; Lin, M. C.; Chen, H. M.; Wang, C. J. Mulberry Leaf Extract Inhibits Vascular Smooth Muscle Cell Migration Involving A Block Of Small GTPase and Akt/NF-kappaB Signals. *J. Agric. Food Chem.* **2009,** *57* (19), 9147–9153.

11. Chen, P. N.; Chu, S. C.; Chiou, H. L.; Kuo, W. H.; Chiang, C. L.; Hsieh, Y. S. Mulberry Anthocyanins, Cyanidin 3-Rutinoside And Cyanidin 3-Glucoside, Exhibited An Inhibitory Effect On The Migration And Invasion Of A Human Lung Cancer Cell Line. *Cancer Lett.* **2006,** *235* (2), 248–259.

12. Cheng, G.; Gao, F.; Sun, X.; Bi, H.; Zhu, Y. Paris Saponin VII Suppresses Osteosarcoma Cell Migration And Invasion By Inhibiting MMP-2/9 Production Via The p38 MAPK Signaling Pathway. *Mol. Med. Rep.* **2016,** *14* (4), 3199–3205.

13. Chompoo, J.; Upadhyay, A.; Fukuta, M.; Tawata, S. Effect of *Alpinia Zerumbet* Components On Antioxidant And Skin Diseases-Related Enzymes. *BMC Complement. Altern. Med.* **2012,** *12* (1), 106.

14. Fraser, J. R. E.; Laurent, T. C.; Laurent, U. B. G. Hyaluronan: Its Nature, Distribution, Functions And Turnover. *J. Intern. Med.* **1997,** *242* (1), 27–33.

15. Garg, C.; Khurana, P.; Garg, M. Molecular Mechanisms Of Skin Photoaging And Plant Inhibitors. *Int. J. Green Pharm.* **2017,** *11* (2), S217–S232.

16. Gopi, K.; Anbarasu, K.; Renu, K.; Jayanthi, S. Quercetin-3-O-Rhamnoside From *Euphorbia Hirta* Protects Against Snake Venom Induced Toxicity. *Biochim. Biophys. Acta* **2016,** *1860* (7), 1528–1540.

17. Grabowska, K.; Podolak, I.; Galanty, A. *In Vitro* Anti-Denaturation And Anti-Hyaluronidase Activities Of Extracts And Galactolipids From Leaves Of *Impatiens Parviflora* DC. *Nat. Prod. Res.* **2016,** *30* (10), 1219–1223.

18. Huang, Y. W.; Chuang, C. Y.; Hsieh, Y. S. *Rubus Idaeus* Extract Suppresses Migration And Invasion Of Human Oral Cancer By Inhibiting MMP-2 Through Modulation Of The Erk1/2 Signaling Pathway. *Environ. Toxicol.* **2017,** *32* (3), 1037–1046.

19. Ishiwa, J.; Sato, T.; Mimaki, Y.; Sashida, Y.; Yano, M.; Ito, A. A Citrus Flavonoid, Nobiletin, Suppresses Production And Gene Expression of MMM-9/Gelatinase B in Rabbit Synovial Fibroblasts. *J. Rheumatol.* **2000,** *27* (1), 20–25.

20. Kim, H. M.; Bae, S. J.; Kim, D. W.; Kim, B. K. Inhibitory Role Of Magnolol On Proliferative Capacity And Matrix Metalloproteinase-9 Expression in TNF-Alpha-Induced Vascular Smooth Muscle Cells. *Int. Immunopharmacol.* **2007,** *7* (8), 1083–1091.

21. Kim, J. H.; Jeong, J. H.; Jeon, S. T.; Kim, H. Decursin Inhibits Induction Of Inflammatory Mediators By Blocking Nuclear Factor-KappaB Activation In Macrophages. *Mol. Pharmacol.* **2006,** *69* (6), 1783–1790.

22. Kim, J.; Kang, Y. R.; Thapa, D.; Lee, J. S. Anti-Invasive And Anti-Angiogenic Effects Of Xanthohumol And Its Synthetic Derivatives. *Biomol. Ther.* **2009,** *17* (4), 422–429.

23. Kim, M. M.; Van Ta, Q.; Mendis, E. Phlorotannins In *Ecklonia Cava* Extract Inhibit MMP Activity. *Life Sci.* **2006,** *79* (15), 1436–1443.

24. Kong, C. S.; Kim, Y. A.; Kim, M. M.; Park, J. S. Flavonoid Glycosides Isolated From *Salicornia Herbacea* Inhibit Matrix Metalloproteinase In HT1080 Cells. *Toxicol. Vitro* **2008,** *22* (7), 1742–1748.

25. Langevin, S. M.; Koestler, D. C.; Christensen, B. C. Peripheral Blood DNA Methylation Profiles Are Indicative Of Head And Neck Squamous Cell Carcinoma: An Epigenome-Wide Association Study. *Epigenetics* **2012,** *7* (3), 291–299.

26. Lee, K. J.; Hwang, S. J.; Choi, J. H.; Jeong, H. G. Saponins Derived From The Roots Of *Platycodon Grandiflorum* Inhibit HT-1080 Cell Invasion And MMPs Activities: Regulation of NF-κB Activation Via ROS Signal Pathway. *Cancer Lett.* **2008,** *268* (2), 233–243.

27. Lee, K.; Chen, Y. S.; Judson, J. P. The Effect Of Water Extracts Of *Euphorbia Hirta* On Cartilage Degeneration In Arthritic Rats. *Malays. J. Pathol.* **2008,** *30* (2), 95–102.

28. Lee, S. J.; Cho, Y. H.; Park, K.; Kim, E. J. Inhibitory Effects Of The Aqueous Extract Of *Magnolia Officinalis* On The Responses Of Human Urinary Bladder Cancer 5637 Cells *In Vitro* And Mouse Urinary Bladder Tumors Induced By N-Butyl-N-(4-hydroxybutyl) Nitrosamine *In Vivo. Phytother. Res.* **2009,** *23* (1), 20–27.

29. Lee, S. K.; Chun, H. K.; Yang, J. Y.; Han, D. C. Inhibitory Effect Of Obovatal On The Migration And Invasion Of HT1080 Cells Via The Inhibition Of MMP-2. *Bioorg. Med. Chem.* **2007,** *15* (12), 4085–4090.

30. Lee, S. O.; Jeong, Y. J.; Kim, M.; Kim, C. H.; Lee, I. S. Suppression Of PMA-Induced Tumor Cell Invasion By Capillarisin Via The Inhibition Of NF-κB-Dependent MMP-9 Expression. *Biochem. Biophys. Res. Commun.* **2008,** *366* (4), 1019–1024.

31. Lee, T. H.; Hsu, C. C.; Hsiao, G.; Fang, J. Y. Anti-MMP-2 Activity And Skin-Penetrating Capability Of The Chemical Constituents From *Rhodiola Rosea. Planta Med.* **2016,** *82* (8), 698–704.

32. Li, L. H.; Baek, I. K.; Kim, J. H.; Kang, K. H. Methanol Extract Of *Elaeagnus Glabra*, A Korean Medicinal Plant, Inhibits HT1080 Tumor Cell Invasion. *Oncol. Rep.* **2009,** *21* (2), 559–563.

33. Liang, Q.; Chen, H.; Zhou, X.; Deng, Q.; Hu, E. Optimized Microwave Assistant Extraction Combined Ultrasonic Pretreatment Of Flavonoids From *Periploca Forrestii* Schltr. And Evaluation Of Its Anti-Allergic Activity. *Electrophoresis* **2017,** *38* (8), 11131121.

34. Limtrakul, P.; Yodkeeree, S.; Thippraphan, P.; Punfa, W.; Srisomboon, J. Anti-Aging And Tyrosinase Inhibition Effects Of *Cassia Fistula* Flower Butanolic Extract. *BMC Complement. Altern. Med.* **2016,** *16* (1), 497.

35. Liotta, L. A.; Tryggvason, K.; Garbisa, S.; Hart, I.; Foltz, C. M.; Shafie, S. Metastatic Potential Correlates With Enzymatic Degradation Of Basement Membrane Collagen. *Nature* **1980,** *284* (5751), 67–68.

36. Liu, C. T.; Bi, K. W.; Huang, C. C.; Wu, H. T.; Ho, H. Y.; Pang, J. H. S.; Huang, S. T. *Davallia Bilabiata* Exhibits Anti-Angiogenic Effect With Modified MMP-2/TIMP-2 Secretion And Inhibited VEGF Ligand/Receptors Expression In Vascular Endothelial Cells. *J. Ethnopharmacol.* **2017,** *196,* 213–224.

37. Liu, F.; Zhang, J.; Yu, S.; Wang, R.; Wang, B.; Lai, L.; Yin, H.; Liu, G. Inhibitory Effect Of *Ginkgo Biloba* Extract On Hyperhomocysteinemia-Induced Intimal Thickening In Rabbit Abdominal Aorta After Balloon Injury. *Phytother. Res.* **2008,** *22* (4), 506–510.

38. Man, S.; Gao, W.; Zhang, Y.; Yan, L. Antitumor And Antimetastatic Activities Of Rhizoma Paridis Saponins. *Steroids* **2009,** *74* (13), 1051–1056.

39. Manosroi, J.; Chankhampan, C.; Kumguan, K.; Manosroi, W.; Manosroi, A. *In Vitro* Anti-Aging Activities Of Extracts From Leaves Of Ma Kiang (*Cleistocalyx Nervosum* Var. *Paniala*). *Pharmaceut. Biol.* **2015,** *53* (6), 862–869.

40. Marquina, M. A.; Corao, G. M.; Araujo, L.; Buitrago, D.; Sosa, M. Hyaluronidase Inhibitory Activity From The Polyphenols In The Fruit Of Blackberry (*Rubus Fruticosus* B.). *Fitoterapia* **2002,** *73* (7), 727–729.

41. McAtee, C. O.; Barycki, J. J.; Simpson, M. A. Emerging Roles for Hyaluronidase in Cancer Metastasis and Therapy. In *Advances in Cancer Research*; Simpson, M. A. and Heldin, P., Eds.; Academic Press: New York, USA, 2014; Vol. 123; pp 1–34.

42. Meenakshi, J.; Shanmugam, G. Inhibition of Matrix Metalloproteinase-9 Activity By Cleistanthin-A, A Diphyllin Glycoside From *Cleistanthus Collinus. Drug Dev. Res.* **2000,** *50* (2), 193–194.

43. Michael, P.; Owczarek, A.; Matczak, M.; Kosno, M. Metabolite Profiling Of Eastern Teaberry (*Gaultheria Procumbens* L.) Lipophilic Leaf Extracts With Hyaluronidase And Lipoxygenase Inhibitory Activity. *Molecules* **2017,** *22* (3), 412.

44. Miláčková, I.; Kapustová, K.; Mučaji, P.; Hošek, J. Artichoke Leaf Extract Inhibits AKR1B1 and Reduces NF-κB Activity In Human Leukemic Cells. *Phytother. Res.* **2017,** *31* (3), 488–496.

45. Mohieldin, E. A. M.; Muddathir, A. M.; Mitsunaga, T. Inhibitory Activities Of Selected Sudanese Medicinal Plants On *Porphyromonas Gingivalis* And MMP-9 And Isolation Of Bioactive Compounds From *Combretum Hartmannianum* (Schweinf) Bark. *BMC Complement. Altern. Med.* **2017,** *17* (1), 224.

46. Moon, H. I.; Cho, H. S.; Kim, E. K. Identification Of Potential And Selective Collagenase, Gelatinase Inhibitors From *Crataegus Pinnatifida. Bioorg. Med. Chem. Lett.* **2010,** *20* (3), 991–993.

47. Moon, H. I.; Chung, J. H. Meso-Dihydroguaiaretic Acid From *Machilus Thunbergii* And Its Effects On The Expression Of Matrix Metalloproteinase-2,9 Caused By Ultraviolet Irradiated Cultured Human Keratinocyte Cells (HaCaT). *Biol. Pharm. Bull.* **2005,** *28* (11), 2176–2179.

48. Mukherjee, P. K.; Maity, N.; Nema, N. K.; Sarkar, B. K. Natural Matrix Metalloproteinase Inhibitors: Leads From Herbal Resources. In *Studies in Natural Products Chemistry*; Rahman, A. U., Ed.; Elsevier: Amsterdam-Netherlands, 2013; Vol. 39; pp 91–111.

49. Murata, T.; Sasaki, K.; Sato, K.; Yoshizaki, F. Matrix Metalloproteinase-2 Inhibitors From *Clinopodium Chinense* Var. *Parviflorum. J. Nat. Prod.* **2009,** *72* (8), 1379–1384.

50. Nelson, A. R.; Fingleton, B. Matrix Metalloproteinases: Biologic Activity And Clinical Implications. *J. Clin. Oncol.* **2000,** *18,* 1135–1149.

51. Olson, M.W.; Toth, M.; Gervasi, D.C.; Sado, Y.; Ninomiya, Y.; Fridman, R. High Affinity Binding Of Latent Matrix Metalloproteinase-9 To The A2(IV) Chain Of Collagen IV. *J. Biol. Chem.* **1998,** *273,* 10672–10681.

52. Opdenakker, G.; Van Damme, J. Cytokine-Regulated Proteases In Autoimmune Diseases. *Immunol. Today* **1994,** *15* (3), 103–107.

53. Orlando, Z.; Lengers, I.; Melzig, M. F.; Buschauer, A.; Hensel, A.; Jose, J. Autodisplay Of Human Hyaluronidase Hyal-1 On *Escherichia Coli* And Identification Of Plant-Derived Enzyme Inhibitors. *Molecules* **2015,** *20* (9), 15449–15468.

54. Ou, Y.; Li, W.; Li, X.; Lin, Z.; Li, M. Sinomenine Reduces Invasion And Migration Ability In Fibroblast-Like Synoviocytes Cells Co-Cultured With Activated Human Monocytic THP-1 Cells By Inhibiting The Expression Of MMP-2, MMP-9, CD147. *Rheumatol. Int.* **2011,** *31* (11), 1479–1485.

55. Park, W. H.; Kim, S. H.; Kim, C. H. New Matrix Metalloproteinase-9 Inhibitor 3, 4-Dihydroxycinnamic Acid (Caffeic Acid) From Methanol Extract Of *Euonymus Alatus*: Isolation And Structure Determination. *Toxicology* **2005,** *207* (3), 383–390.

56. Philips, N.; Conte, J.; Chen, Y. J.; Natrajan, P. Beneficial Regulation Of Matrixmetalloproteinases And Their Inhibitors, Fibrillar Collagens And Transforming Growth Factor-Beta By *Polypodium Leucotomos*, Directly Or In Dermal Fibroblasts, Ultra-Violet Radiated Fibroblasts, And Melanoma Cells. *Arch. Dermatol. Res.* **2009,** *301* (7), 487–495.

57. Piwowarski, J. P.; Kiss, A. K.; Kozłowska-Wojciechowska, M. Anti-Hyaluronidase And Anti-Elastase Activity Screening Of Tannin-Rich Plant Materials Used In Traditional Polish Medicine For External Treatment Of Diseases With Inflammatory Background. *J. Ethnopharmacol.* **2011,** *137* (1), 937–941.

58. Radhakrishnan, A. I.; Wai, L. K.; Ismail, I. S. Molecular Docking Analysis Of Selected *Clinacanthus Nutans* Constituents As Xanthine Oxidase, Nitric Oxide Synthase, Human Neutrophil Elastase, MPP-2, MMP-9 And Squalene Synthase Inhibitors. *Pharmacogn. Mag.* **2016,** *12* (Suppl. 1), S21–S26.

59. Rahbardar, M. G.; Amin, B.; Mehri, S. Anti-Inflammatory Effects Of Ethanolic Extract Of *Rosmarinus Officinalis* L. And Rosmarinic Acid In A Rat Model Of Neuropathic Pain. *Biomed. Pharmacother.* **2017,** *86*, 441–449.

60. Rasouli, H.; Parvaneh, S.; Mahnam, A.; Rastegari-Pouyani, M.; Hoseinkhani, Z.; Mansouri, K. Anti-Angiogenic Potential Of Trypsin Inhibitor Purified From *Cucumis Melo* Seeds: Homology Modeling And Molecular Docking Perspective. *Int. J. Biol. Macromol.* **2017,** *96*, 118–128.

61. Rodrigues, E. R.; Nogueira, N. G. P.; Zocolo, G. J. Pothomorphe Umbellata: Antifungal Activity Against Strains Of *Trichophyton Rubrum. J. Mycol. Méd.* **2012,** *22* (3), 265–269.

62. Rzany, B.; Becker-Wegerich, P.; Bachmann, F.; Erdmann, R.; Wollina, U. Hyaluronidase In The Correction Of Hyaluronic Acid-Based Fillers: A Review And A Recommendation For Use. *J. Cosmet. Dermatol.* **2009,** *8* (4), 317–323.

63. Salvamani, S.; Gunasekaran, B.; Shukor, M. Y.; Shaharuddin, N. A. Anti-HMG-CoA Reductase, Antioxidant, And Anti-Inflammatory Activities Of *Amaranthus Viridis* Leaf Extract As A Potential Treatment For Hypercholesterolemia. *Evid. Based Complement. Altern. Med.* **2016,** *2016*, 10.

64. Scotti, L.; Kumar Singla, R.; Mitsugu Ishiki, H. Recent Advancement In Natural Hyaluronidase Inhibitors. *Curr. Top. Med. Chem.* **2016,** *16* (23), 2525–2531.

65. Shan, Y. F.; Shen, X.; Xie, Y. K.; Chen, J. C.; Shi, H. Q.; Yu, Z. P. Inhibitory Effects Of Tanshinone II-A On Invasion And Metastasis Of Human Colon Carcinoma Cells. *Acta Pharmacol. Sin.* **2009,** *30* (11), 1537–1542.

66. Sternlicht, M. D.; Werb, Z. How Matrix Metalloproteinases Regulate Cell Behavior. *Ann. Rev. Cell Dev. Biol.* **2001,** *17* (1), 463–516.

67. Suh, S. J.; Cho, K. J.; Moon, T. C.; Chang, H. W.; Park, Y. G.; Kim, C. H. 3, 4, 5-Trihydroxybenzaldehyde From *Geum Japonicum* Has Dual Inhibitory Effect On MMP-9: Inhibition Of Gelatinoytic Activity As Well As MMP-9 Expression In TNF-α Induced HASMC. *J. Cell. Biochem.* **2008,** *105* (2), 524–533.

68. Tanimura, S.; Kadomoto, R.; Tanaka, T.; Zhang, Y. J.; Kouno, I.; Kohno, M. Suppression Of Tumor Cell Invasiveness By Hydrolyzable Tannins (Plant Polyphenols) Via The Inhibition Of Matrix Metalloproteinase-2/-9 Activity. *Biochem. Biophys. Res. Commun.* **2005,** *330* (4), 1306–1313.

69. Tayeh, M.; Nilwarangoon, S.; Mahabusarakum, W. Anti-Metastatic Effect Of Rhodomyrtone From *Rhodomyrtus Tomentosa* On Human Skin Cancer Cells. *Int. J. Oncol.* **2017,** *50* (3), 1035–1043.

70. Tomohara, K.; Ito, T.; Onikata, S.; Kato, A.; Adachi, I. Discovery Of Hyaluronidase Inhibitors From Natural Products And Their Mechanistic Characterization Under DMSO-Perturbed Assay Conditions. *Bioorg. Med. Chem. Lett.* **2017,** *27* (7), 1620–1623.

71. Torres-Carro, R.; Isla, M. I.; Thomas-Valdes, S. Inhibition Of Pro-Inflammatory Enzymes By Medicinal Plants From The Argentinean Highlands (Puna). *J. Ethnopharmacol.* **2017,** *205*, 57–68.

72. Yang, S. F.; Chu, S. C.; Liu, S. J.; Chen, Y. C. Antimetastatic Activities Of *Selaginella Tamariscina* (Beauv.) On Lung Cancer Cells *In Vitro* And *In Vivo*. *J. Ethnopharmacol.* **2007,** *110* (3), 483–489.

73. Yanti, O. H. I.; Anggakusuma, H. J. K. Effects Of Panduratin A Isolated From *Kaempferia Pandurata* ROXB. On The Expression Of MMP-9 By *Porphyromonas Gingivalis* Supernatant-Induced KB Cells. *Biol. Pharm. Bull.* **2009,** *32* (1), 110–115.

74. Yoon, S. O.; Shin, S.; Lee, H. J.; Chun, H. K.; Chung, A. S. Isoginkgetin Inhibits Tumor Cell Invasion By Regulating Phosphatidylinositol 3-Kinase/Akt–Dependent MMP-9 Expression. *Mol. Cancer Ther.* **2006,** *5* (11), 2666–2675.

75. Zhou, H.; Wong, Y.F.; Wang, J.; Cai, X.; Liu, L. Sinomenine Ameliorates Arthritis Via MMPs, TIMPs And Cytokines In Rats. *Biochem. Biophys. Res. Commun.* **2008,** *376* (2), 352–357.

76. Zhu, G. H.; Dai, H. P.; Shen, Q.; Ji, O.; Zhang, Q.; Zhai, Y. L. Curcumin Induces Apoptosis And Suppresses Invasion Through MAPK And MMP Signaling In Human Monocytic Leukemia SHI-1 Cells. *Pharmaceut. Biol.* **2016,** *54* (8), 1303–1311.

INDEX

Printed and bound by CPI Group (UK) Ltd, Croydon, CR0 4YY

23/10/2024

01777702-0011